Environmental Footprints and Eco-design of Products and Processes

Series Editor

Subramanian Senthilkannan Muthu, Head of Sustainability - SgT Group and API, Kowloon, Hong Kong

This series aims to broadly cover all the aspects related to environmental assessment of products, development of environmental and ecological indicators and eco-design of various products and processes. Below are the areas fall under the aims and scope of this series, but not limited to: Environmental Life Cycle Assessment; Social Life Cycle Assessment; Organizational and Product Carbon Footprints; Ecological, Energy and Water Footprints; Life cycle costing; Environmental and sustainable indicators; Environmental impact assessment methods and tools; Eco-design (sustainable design) aspects and tools; Biodegradation studies; Recycling; Solid waste management; Environmental and social audits; Green Purchasing and tools; Product environmental footprints; Environmental management standards and regulations; Eco-labels; Green Claims and green washing; Assessment of sustainability aspects.

More information about this series at http://www.springer.com/series/13340

Subramanian Senthilkannan Muthu
Editor

Water Footprint

Assessment and Case Studies

 Springer

Editor
Subramanian Senthilkannan Muthu
Head of Sustainability
SgT Group and API
Kowloon, Hong Kong

ISSN 2345-7651 ISSN 2345-766X (electronic)
Environmental Footprints and Eco-design of Products and Processes
ISBN 978-981-33-4379-5 ISBN 978-981-33-4377-1 (eBook)
https://doi.org/10.1007/978-981-33-4377-1

This Springer imprint is published by the registered company Springer Nature Singapore Pte Ltd.
The registered company address is: 152 Beach Road, #21-01/04 Gateway East, Singapore 189721, Singapore

This book is dedicated to

The lotus feet of my beloved Lord Pazhaniandavar

My beloved late Father

My beloved Mother

My beloved Wife Karpagam and Daughters— Anu and Karthika

My beloved Brother—Raghavan

Everyone working in various industrial sectors to reduce the water footprint to make our planet earth SUSTAINABLE

Contents

About the Editor

Dr. Subramanian Senthilkannan Muthu currently works for SgT Group as Head of Sustainability, and is based out of Hong Kong. He earned his Ph.D. from The Hong Kong Polytechnic University, and is a renowned expert in the areas of Environmental Sustainability in Textiles & Clothing Supply Chain, Product Life Cycle Assessment (LCA) and Product Carbon Footprint Assessment (PCF) in various industrial sectors. He has five years of industrial experience in textile manufacturing, research and development and textile testing and seven years of experience in life cycle assessment (LCA), carbon and ecological footprints assessment of various consumer products. He has published more than 75 research publications, written numerous book chapters and authored/edited over 85 books in the areas of Carbon Footprint, Recycling, Environmental Assessment and Environmental Sustainability.

Verifiable Water Use Inventory Using ICTs in Industrial Agriculture

Carmen M. Flores-Cayuela, Rafael González-Perea, Emilio Camacho-Poyato, and Pilar Montesinos

Abstract Since the 1950s, the world's population has increased by more than six billion people. This exponential growth led to changes in the irrigated agricultural sector, the world's largest user of water resources (70%), focusing on increasing food production to meet the growing needs of the population. With the green revolution, "industrial agriculture" appeared, transferring some of the industrial production principles to the farming sector in order to reduce year-to-year variability in harvests and increase production on a sustained basis. This has led to the specialization of farms in a single product (monoculture), the technification and intensification of these farms and the increase in the use of inputs. The digitalization of the agricultural sector, through the implementation of cutting-edge technologies, allows for the optimization of the use of key resources such as water, whose scarcity has become one of the most relevant and complex environmental global problems. Water consumption data collected on site, using different technologies, provides a real and transparent inventory of water use in irrigated crops. The water footprint obtained in accordance with the ISO 14046 standard (within the LCA methodology), jointly with the irrigation management information, allows knowing accurately the efficiency in the use of water of each crop production system. The installation of different devices to monitor water use in greenhouse tomatoes and citrus groves has made it possible to collect and quantify precisely the water used by these crops throughout the irrigation season.

C. M. Flores-Cayuela (✉) · E. Camacho-Poyato · P. Montesinos
Department of Agronomy, University of Córdoba, Campus Rabanales, Edif. Da Vinci, 14071
Córdoba, Spain
e-mail: g02flcac@uco.es

E. Camacho-Poyato
e-mail: ecamacho@uco.es

P. Montesinos
e-mail: pmontesinos@uco.es

R. González-Perea
Department of Plant Production and Agricultural Technology, School of Advanced Agricultural
Engineering, University of Castilla-La Mancha, Campus Universitario, s/n, 0207 Albacete, Spain
e-mail: Rafael.GonzalezPerea@uclm.es

Having a real and verifiable water inventory during the growing season is essential to carry out a correct water footprint assessment and to establish strategies that allow better water management and efficient use of water, reducing its environmental impact at the farm level.

Keywords Water footprint inventory · Information and communication technologies · Wireless sensors · Irrigation management · Green water use · Blue water use · Efficient water use

List of Acronyms

A	Irrigation sector area (ha)
ADL	Allowable depletion level (mm)
API	Application Programming Interface
CWA_{blue}	Irrigation water applied to the crop (m^3/ha)
CWU_{blue}	Irrigation water evapotranspirated by crop (m^3/ha)
CWU_{green}	Green component of the crop water use (m^3/ha)
DOY	Day of the year
Dr	Soil moisture deficit (mm)
ER	Effective rainfall (mm)
ET_{blue}	Blue water use by the crop (m^3/ha)
ETC	Crop evapotranspiration (mm)
ET_{green}	Green water use by the crop (m^3/ha)
ET_0	Reference evapotranspiration (mm)
FC	Field capacity
FIR	Full irrigation requirements strategy
ICTs	Information and communication technologies
I	Real irrigation sheet apply and measure with water meter (mm)
IDSS	Irrigation decision support system
IE	Irrigation system efficiency (%)
IN	Theoretical irrigation needs
IR	Real irrigation requirements
IWA	Irrigation water allowance (mm)
K_C	Crop coefficient
K_s	Non-dimensional coefficient of water stress in the soil
LCA	Life cycle analysis
lds	Last day of the season
n	Index day of the season
n_e	Number of emitters per sector
NAE	Nearest agroclimatic station
θ	Volumetric water content (m^3/m^3)
θ_{FC}	Volumetric water content at field capacity (m^3/m^3)
θ_{PWP}	Volumetric water content at permanent wilting point (m^3/m^3)

P	Gross precipitation (mm)
PWP	Permanent wilting point
q_e	Emitters flown (L/h)
RAW	Readily available water (mm)
RDI	Regulated deficit irrigation strategy
RIS	Relative water supply
Rs_{in}	Solar radiation inside greenhouse (mm)
Rs_{out}	Outdoor solar radiation (mm)
SDI	Sustained deficit irrigation strategy
TAW	Total available water (mm)
t_t	Theoretical irrigation time (hours)
t_r	Real irrigation time (hours)
WA_g	Irrigation water applied during the crop growing (m³/ha)
WA_p	Amount of water use in different crop management practices (m³/ha)
WFA_{blue}	Blue water footprint applied (m³/t)
WF_{blue}	Blue component of the water footprint of the crop (m³/t)
WF_C	Crop water footprint (m³/t)
WF_{green}	Green component of the water footprint of the crop (m³/t)
WFN	Water footprint network
wm	Number of water meters installed in a farm
Y	Crop yield (t/ha)
Zr	Root depth (m)

1 Introduction

Water has a fundamental role in the development of humanity. Economic growth, poverty reduction, human health and environmental conservation are closely linked to water resource use [51].

Increasing demand for freshwater has become one of the world's greatest and most complex challenges [31]. It is necessary to restore the balance between the demand and the increasingly limited supply of water resources. If this balance is not achieved, the world will face a situation of growing water scarcity that could affect more than two-thirds of the world's population by 2025 [50]. In addition to the natural factors which affect water resource availability, human activity has become the main driver of water pressures. Freshwater demand is driven by different factors such as world population growth, climate change, water policies and legislation, technological innovation processes and socioeconomic factors [50].

Irrigated agriculture is the largest human activity using freshwater resources. 70% of the world's freshwater withdrawals are used for agricultural production [13]. The demand for freshwater in the agricultural sector is driven by the demand for food to meet the population's needs.

In 1950, the world population was 2536 million, reaching 7794 million today. It took one million years to attain one billion people in the world (around 1820), one hundred and ten years to double that figure (1930), forty for three billion (1960), fifteen (1975) and twelve years (1987) to reach four and five billion, respectively [46]. The exponential growth experienced by the world's population, especially since the 1950s, has prompted the development of new crop farming systems to increase food production. The world population is expected to reach 8.6 billion by 2030 and 9.8 billion by 2050, so global water demand will increase by 20–30% by 2050 [52]. Agriculture, the main user of water resources, will have to take up the challenge of increasing global food production by up to 60% by 2050 in a context of limited resources [51].

Although the extension of agricultural land has increased, it is insufficient to satisfy the growing demand for agricultural products, making it necessary to increase crop yields [44]. The so-called "Green Revolution" (between 1960 and 1980) gave way to a new, more industrialized agricultural production system (industrial agriculture) characterized by the specialization of farms in a single product (monoculture), the technification and intensification of these farms and the increase in production associated with greater capital investment and higher use of inputs: fertilizers, herbicides, machinery, irrigation infrastructure, genetic selection of seeds, development of new high-yield varieties, etc. [46].

Climate change is another key element of pressure on water resources in terms of quantity and quality. There is evidence that the world's climate is changing and the main impact of this change on humans is through water. Increased frequency of extreme events (droughts and floods), decreased river flow, increased torrential rainfall and rising temperatures (leading to higher crop water demand) are some of the consequences of climate change on water resources [50]. Thus, the growing use of inputs brought about by Green Revolution, together with the climate change, have intensified the negative effects on the quantity and quality of water resources, such as salinization of irrigated areas, pollution of water bodies with nitrates from the agricultural activity and overexploitation of aquifers [13].

World agriculture must face a huge challenge: to increase food production to meet the growing food demand without compromising the preservation of the environment, in a scenario of reduced water availability, more frequent droughts and uncertainty associated with climate change. Irrigated agriculture must produce more with less. This means that water productivity must be increased by improving water use efficiency [41].

Installation of high-efficiency irrigation systems for water application (e.g. localized irrigation) and the digitalization of the agricultural sector, through the application of cutting-edge technologies, is one of the strategies proposed to make more efficient use of water resources. In this context, the implementation of precision irrigation systems, based on efficient hydraulic installations, use of sensors and information and communication technologies (ICT's) arises as a possible solution to increase productivity and reduce the environmental impact of irrigated agriculture [41]. Precision irrigation enables optimal water use while maintaining or even increasing crop production in quantity and quality. This implies precise knowledge of crop water needs, optimal irrigation programming for the application of water at the right time, and the use of high-efficiency hydraulic elements that allow for uniform spatial application [15, 49].

From this point of view, drip irrigation systems are an ideal option to achieve more efficient use of water, especially in arid regions where water resources are scarce and expensive. Drip irrigation systems apply water in a localized manner with an efficiency of more than 90%. They wet only the soil fraction occupied by plant roots and maintain an optimal moisture level. In addition, the subdivision of the irrigation network into sectors adapted to the spatial variability of the farms allows for flexible and variable water applications. These systems are easy to manage due to their high degree of automation, and together with the abovementioned characteristics make these irrigation systems the core of precision irrigation systems [11, 21, 22, 41, 56].

To adapt the amount and timing of irrigation to crop irrigation needs, it is necessary to know both the spatial and temporal variations in soil moisture levels and crop water needs during their development stages. Until recently, irrigation management at the farm level was generally based on the farmer's experience without a scientific basis. From this perspective, the development of ICTs allows the implementation of efficient water use in the irrigated agriculture sector. Information collected in the field through remote sensors and sent to the analysis and decision-making center using ICTs is a critical element for the implementation of precision irrigation systems. Despite recent advances in the field of wireless sensor networks and mobile communication systems that allow control and monitoring of crop status in real time, they only provide punctual measurements of water availability or crop needs at a specific point at the plot scale. Moreover, this information is usually analyzed individually, and it is "the irrigator" who combines the information recorded by the sensor with his own knowledge about the plot characteristics to determine the duration and timing of irrigation.

These advances have boosted the development of ICT applications in agriculture, especially in irrigation management [20–23, 26, 43, 54, 57]. Some of these works have been focused on the development of computer tools to support farmers in the correct management of their water resources. González Perea et al. [26] developed a mobile and desktop application that uses ICT tools to determine the daily irrigation time for strawberry crops in southwestern Spain, using agroclimatic and soil information as well as the hydraulics of the irrigation network. The application developed by Alcaide Zaragoza et al. [1] determines in real time the optimal fertigation schedule for olive groves, taking into account the quality of water used for irrigation. Using ICTs, it combines climate forecasting with agroclimatic records to determine crop water needs in real time. The algorithms developed in this work establish the timing, quantity and frequency of irrigation/fertigation based on real-time agroclimatic information, weather forecast, crop and soil characteristics and theoretical estimates of the water content available in the soil for the crop. However, the application of variable irrigation volumes, adapted to the heterogeneity of the farm, requires data collected in situ and in real time by a network of permanent sensors [41].

Water crisis is not only an environmental problem, but also an economic one. According to the World Economic Forum's global risk report [55], the failure to mitigate and adapt to climate change and the occurrence of extreme weather events are among the three main risks both in terms of probability and impact. Thus, all actors involved in water use (water policy makers, water supply companies, managers, non-governmental organizations and consumers) are looking for more sustainable ways to manage it [18]. As irrigated agriculture is the largest user of water resources, it is crucial that farmers adopt strategies aimed at sustainable use of irrigation water. In this context, the water use indicator called water footprint (WF) emerges as a tool for evaluating these strategies. Hoekstra [29] introduced the concept of WF by defining it as the volume of water used, directly or indirectly, to produce a unit of output (product or service). Since then the concept of WF has been in constant evolution. Several authors have included WF in life cycle analysis [4, 6, 7, 10, 33, 45]. It is in 2014 when the standard ISO 14046:2014—Environmental Management—Water Footprint—Principles, requirements and guidelines [32]—was published in response to the need for unifying the multiple definitions and methodologies of Water Footprint assessment.

ISO 14046:2014 defines the water footprint as a metric or metrics that quantify potential environmental impacts related to water. This methodology is based on the life cycle analysis (LCA) approach. It collects and evaluates inputs, outputs and potential environmental impacts related to water "from cradle to grave".

According to ISO 14046, Water Footprint assessment has 4 steps: (i) objective and scope definition, (ii) water footprint inventory analysis, (iii) water footprint impact assessment and (iv) finally, result interpretation. Activities related to the first and second phases are of great significance. Clearly, defining the spatial-temporal scale affects the accuracy of the WF inventory results and their interpretation [25, 42].

In the Water Footprint inventory phase of irrigated crops, water inputs and outputs are collected and quantified. So it is a useful process to evaluate the relationship between water use and crop yields, so water footprinting is a measure of water use efficiency in crops [21, 22]. When the objective is to formulate strategies to improve water use efficiency, it is desirable that the characterization of the water use of the crop under study be based on local information on water availability and use (on-site studies) [25, 34]. However, this type of data is usually not readily available, and average data provided by specialized literature or public databases is generally used. Access to quality data about water use is a factor that conditions the assessments of water use impacts. In most agricultural and industrial processes, real data on water use are scarce due to the small number of farmers and companies that collect or report information about their water use. Getting this information in a meaningful and verifiable format would be a step forward in developing suitable methods for estimating the effects of water consumption [34].

The current development of ICTs and their implementation at the farm level allows the design of accurate procedures to inventory water use at this scale. Such an inventory is a key tool for analyzing the traceability of water use in a transparent manner at the farm level. Communicating verifiable inventories on water use to market would provide "transparent water use-farmers" with differentiation from farmers who do not provide information on where they take their water, how they use it and how they return it to the environment. This difference is likely to result in higher prices for "transparent products", thus encouraging sustainable water use in agricultural production.

In this work, a procedure has been developed to determine the real-time scheduling of a precision irrigation system that optimizes the use of irrigation water in crops produced according to industrial agriculture techniques. Real-time knowledge of water inputs and outputs during the development of the crop based on data received from sensors installed in the field allows obtaining the verifiable inventory of water use at the farm scale using ICTs.

The proposed methodology has been applied to two crops (greenhouse tomatoes and orange trees) grown according to the principles of industrial agriculture during the 2019/2020 irrigation season. The cultivation area is located in the south of Spain, where a particularly unfavorable climate scenario is expected as a result of climate change: with an increase in temperature, a reduction in precipitation and an increase in torrential rainfall. It is therefore essential to optimize the use of scarce water resources in this region. The efficiency of the implemented precision irrigation systems is determined by water footprinting each crop.

After the introduction section, the organization of the contents of this chapter is described below. Section 2 (materials and methods) includes the description of the case studies in which the proposed methodology has been applied, the procedure for applying ICT to determine the volumes of irrigation water (blue water) and rainwater (green water) used by the crops, and the methodology for optimizing water use and ends by describing how to make an accurate inventory of the water footprint of a crop.

Section 3 shows the application of the developed methodology in two case studies, discussing the results obtained and including a comparison with previous water footprinting works of the studied crops. The chapter finishes gathering the main conclusions obtained in this work, as well as possible lines of research in this field.

2 Methodology

2.1 Case Study Description

The developed methodology has been applied to the inventory water use of an orange grove and a tomato greenhouse crop grown in southern Spain (Fig. 1).

These crops are cultivated by applying the principles of organic and industrial agricultural production.

These two crops are quite different in terms of water requirements. Citrus fruit can use rainwater to cover part of their water needs, while tomatoes are grown in greenhouses throughout their cycle and only use irrigation water to satisfy their needs. Both crops are cultivated in a water scare.

Thus, tomato, in particular, is grown in the southeast, in one of the areas with the largest concentrations of greenhouses in the world, mainly dedicated to intensive

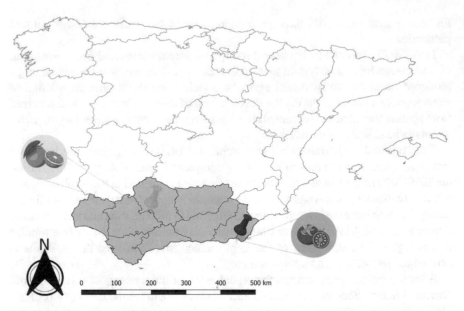

Fig. 1 Location of the study crops

horticultural production [9]. Water resources are very scarce in this region and are severely pressured by intensive agriculture, which aggravates overexploitation, pollution and salinization of aquifers [19]. Due to this situation, there is a growing need to optimize the use of irrigation water, which makes greenhouse tomatoes especially Interesting to test the methodology developed in this chapter.

The greenhouse tomato cycle extends from mid-August to the end of May. The average daily reference evapotranspiration inside the greenhouse during this period is 2.34 mm. The slope of the greenhouse roof, its management (frequency and dose of whitening to reduce the temperature inside the greenhouse), the electrical conductivity of the irrigation water, which change during the irrigation campaign, as well as the characteristics of the irrigation network are aspects that condition the management of tomato irrigation.

In the trial greenhouse, whitening is usually applied from planting to mid-September and the applied dose is 0.25 kg of lime per liter. The average electrical conductivity of the irrigation water is 2 dS/m and the slope of the greenhouse cover is 20°. The soil in the greenhouse has only 27 cm depth (bounded by a hardened profile) and a loam texture. The planting density is 1.5 plants/m². Irrigation water is supplied by an irrigation pond that is fed by own water collection. The irrigation network is composed of irrigation sectors in which each pipe irrigates a line of plants, separated from each other by 2.05 m. The flow rate of the drippers is 4 L/h (not self-compensating) placed every 0.4 m. Under these conditions, the average tomato production is 109 t/ha (branch and cocktail varieties.). The destination of this production is usually the fresh produce markets of Central Europe.

The citrus fruit plantation (Navelina variety) is located in the Guadalquivir valley, an area with greater availability of resources (about 600 mm of average annual rainfall), but with a high percentage of its cultivated area dedicated to irrigated crops (28.92% of total national irrigation) [37]. The high demand for irrigation water in this basin means that the irrigation allocations made by the water authorities must be lower than the water requirements of the crops. It is very frequent for the application of deficit irrigation strategies to make efficient use of the available water.

The average of the average reference evapotranspiration during the orange tree irrigation campaign is 3.65 mm per day. The soil of the farm has a sandy loam texture. The distance between the crop lines is 5.5 m and the distance between the trees is 4 m. The plantation is irrigated from a hydrant of the primary irrigation network of the irrigation district to which the farm belongs. The farm's irrigation network is organized in sectors in which each tree line is irrigated with two irrigation branches, with self-compensating emitters of 2.8 L/h and a separation between them of 1 m.

The mean annual production of the studied orchard is 15 t/ha while being mainly exported to German fresh produce markets.

2.2   Use of ICTs to Measure and Optimize Water Use

Soil moisture sensors and water meters have been installed to monitor crop water use. The capacitive moisture sensors of type FDR (Frequency Domain Reflectometry) were used to monitor and measure changes in soil volumetric water content (VWC).

To ensure that the soil water monitoring devices record variations in soil moisture due to crop absorption, they should be placed in the layer with the largest volume of crop roots. The depth of the roots depends on the crop, its state of development, type of soil, presence of tillage sole, among other factors. After an in situ examination of the soil and root systems of each crop, sensors were installed at 3 levels in increasing order of depth: surface layer, zone with the highest concentration of roots and the depth at which only a small fraction of the roots reach, where high moisture contents would indicate water leakage by percolation and therefore inefficient use of irrigation water.

Water meters record the total volume that has circulated through the pipe in which they were located since their installation. Metering is done by an electronic sensor that emits a pulse with each rotation of a piece moved by the flow of water. The volume of water recorded can be displayed on the meter's analog dial and is then transmitted in digital format to the data server, with the same frequency as soil moisture content records are sent. Both elements are connected to a data logger with General Packet Radio Service (GPRS) communication that records and transmits in real-time data measured by sensors to a remote server. To consult and use the data provided by water meters and soil moisture sensors, it is accessed through an Application Programming Interface (API).

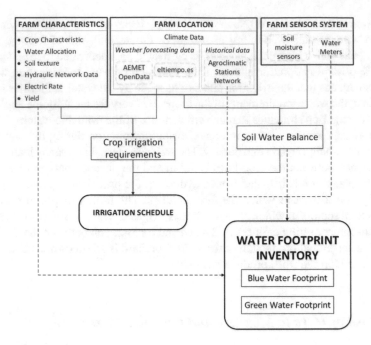

Fig. 2 Scheme of the water footprint inventory process

The developed methodology combines the use of information recorded by sensors installed in the field, with the characteristics of the crop, soil type, hydraulic characteristics of the irrigation network, climate data (both historical and predicted), information on water availability and production, to calculate accurately the water used by the crop in real time in order to build a true and precise water footprint inventory. Simultaneously, this information feeds an irrigation decision support system (IDSS), which facilitates the management of a precision irrigation system to maximize the efficiency of the use of available water resources. This IDSS provides an optimal 7-day irrigation schedule.

Figure 3 shows a scheme of the methodological approach to inventory crop water footprints. The main calculation modules are described below (Fig. 2).

2.2.1 Farm Characteristics

The water footprint inventory methodology is applicable to both crops grown in outdoor conditions as well as greenhouse crops, requiring the flowing data for irrigation scheduling and Water Footprint calculation:

– Crop characteristic: The plantation frame (m m) and average diameter of the tree canopy (m) are required for trees. For greenhouse crops, planting date and spacing between crop lines are required including information about the greenhouse (slope of its cover (°)), its management (start and end dates of cover whitening, whitening dose (kg product/100 L water)) and electrical conductivity of irrigation water (dS/m).

- Characteristics of the irrigation network: number of pipes per crop line, distance between emitters (m), nominal flow rate of the emitters (L/h) and sector area (ha).
- Electric tariff. The electricity rate contracted by each farm conditions the management and use of water when the electricity supplier offers different electricity rates according to the supply voltage and power contracted. Among these tariffs, there are differences in the number of tariff periods and the number of hours and their distribution. Because each tariff period will have different prices of power and energy, for economic or management reasons, the number of hours of irrigation per day is usually limited. To keep the recommended irrigation time within this limit, the user must either select the tariff type and irrigation period or manually enter the maximum number of available irrigation hours from Monday to Friday and on weekends.

Additionally, the water allowance (m^3/(ha year)), soil texture, and dates of harvest end in the last two seasons and production obtained in the last season (t/ha) must be known.

2.2.2 Climate Data

This module uses climate predictions and historical data to estimate crop water needs. In this work, the historical climate records were taken from the Agroclimatic Stations Network of the Regional Government of Andalucía.

The reference evapotranspiration data, ET_0 (mm), recorded at the nearest public agroclimatic station are used to calculate crop evapotranspiration ETC (mm), equivalent to their water requirements. With this information, an internal database is built up with the historical water needs of the crop ETC (mm) and its later use to distribute the available irrigation water for the campaign. For outdoor crops, rainfall records and ETC data will be used to update the daily soil water balance, which is necessary to determine the optimal irrigation timing.

To calculate crop irrigation needs, ET_0 predictions are based on Open weather data (maximum and minimum daily relative humidity information (%)) and on data obtained from online weather forecast (maximum and minimum temperature (°C), cloud index (%) and wind speed (km/h)) by web scraping techniques. These data are used to calculate ET_0 for the next 7 days using the Penman–Monteith equation [3] for outdoor crops.

Plastic covers modify the outdoor ETo, so to estimate its value in the greenhouse the equation proposed by Fernández et al. [17] based on solar radiation in the greenhouse was applied (Eq. 1).

$$ET_{0\,greenhouse} = (0.288 + 0.0019 \cdot DOY) \cdot Rs_{in} \quad \text{for DOY} \leq 220$$
$$ET_{0\,greenhouse} = (1.339 - 0.0028 \cdot DOY) \cdot Rs_{in} \quad \text{for DOY} > 220 \tag{1}$$

where DOY is the Julian Day and Rs_{in} is the solar radiation inside the greenhouse (mm/day) that can be estimated from external radiation data, using the cover transmissivity value (τ) which depends on the material (Eq. 2).

$$Rs_{inv} = Rs_{out} \cdot \tau \tag{2}$$

Rs_{out} is the outdoor solar radiation (mm/day); since climate predictions do not provide this information directly, it has been calculated through Angstrom's method [3], which links solar radiation to extraterrestrial radiation and the relative insolation duration. To estimate this last one, cloudiness predictions (%) are used, which allow an approximation of the real daily insolation that will occur with respect to the maximum possible.

Weather precipitation forecast is considered in the irrigation scheduling of outdoor crops.

2.2.3 Soil Water Content

In outdoor crops, it is necessary to consider the amount of water stored in the soil to determine whether irrigation is necessary or not (when the amount of water stored in the soil is enough to satisfy their ETC). On the contrary, for greenhouse crops with drip irrigation systems and highly frequent irrigation and no rainfall input, it is reasonable to ignore the role of the soil as a water store and consider that the water content in the soil does not change over time [16].

Total water available in the soil (TAW) for the crop is determined by the crop's depth of roots (zr) and by the upper limit θ_{FC} (field capacity) and lower limit θ_{PWP} (permanent wilting point) of soil water storage Eq. (3), which depend on the soil's hydrophilic characteristics, n being the day index.

$$TWA_n = 1000 \cdot (\theta_{FC} - \theta_{PWP}) \cdot zr \tag{3}$$

However, not all the water stored in the soil can be used by the crop, only a fraction of it is really available for the crop (RAW) that varies according to the crop [3]. Once the soil moisture deficit (Dr) reaches the RAW value, water extraction is more difficult for the plant and leads to a reduction of ETc. To avoid this situation, 20% of TAW has been fixed as the threshold of allowable depletion level (ADL) in the soil to start irrigation. To determine the Dr_n (mm) in the soil at the end of each day (soil water balance), the $Dr_{theorical,n}$ and $Dr_{real,n}$ are previously calculated using Eqs. (4) and (5), respectively, adapted from Allen et al. [3]:

$$Dr_{theorical,n} = Dr_{n-1} - ER_n - I_n + (ET_{c_n} \cdot K_{s_n}) \tag{4}$$

Equation (4) estimates theoretical water deficit from the balance of inputs (irrigation and effective rainfall) and outputs (evapotranspiration) at the end of the day, without considering runoff and percolation losses. Dr_{n-1} accounts for the soil moisture depletion of the previous day (mm), ER_n is the effective rainfall of n day (mm) whose calculation is described in the following subsection, I_n is the real applied irrigation depth and is measured with the water meter installed on field (mm), ETC is the crop evapotranspiration (mm) estimated from the ET_0 data registered by the nearest agroclimatic station and information on the state of development of the crop. K_s is a non-dimensional coefficient of water stress [3] that quantifies the reduction of transpiration when water deficit in the soil exceeds RAW. As long as water depletion in the root zone is less than RAW, the value of this coefficient is equal to 1.

Equation (5) calculates the depletion of soil moisture using the information recorded by on-site moisture sensors.

$$Dr_{real,n} = 1000 \cdot (\theta_{FC} - \theta_{real\ n-1}) \tag{5}$$

where θ_{FC} represents the volumetric water content of soil at field capacity (m^3/m^3) and θ_{real} is the average volumetric water content (m^3/m^3) recorded by on-site sensors at a fixed time every day using Eq. (6):

$$\theta_{real} = \frac{\sum_{Z=1}^{p} \theta_z}{n} = \frac{\theta_{Z1} + \theta_{Z2}}{2} \tag{6}$$

where Z identifies the sensor by its depth from the ground surface, so $\theta Z1$ and $\theta Z2$ represent the volumetric water contents recorded by the most superficial sensor and the one located underneath, in the layer with higher root density, respectively. As commented above, the function of the sensor located below Z2 is aimed at controlling percolation occurrence.

To estimate soil water content, θ, the soil water retention curves are needed. These curves are estimated from soil texture. Soil samples taken at sensor location depths will be analyzed to determine soil texture using the US Department of Agriculture (USDA) soil texture triangle. Then the soil water retention curves can be calculated by the ROSETTA model [47]. The retention curves are used to estimate θ_{FC} and θ_{PWP} while $\theta Z1$ and $\theta Z2$ are obtained from the sensors at sunset giving information about the remaining water content in the soil after the daily evapotranspiration period.

Frequently, after irrigation, the water content in the soil exceeds FC for a while, then the soil drains and the water content stabilizes. After irrigation, soil moisture data recorded has high θ values as the soil has not drained yet and the crop has not taken the water available yet. If this data is used, inaccurate irrigation scheduling could be done. It is possible to change this parameter to suit other irrigation and crop management criteria.

If the average moisture value exceeds a certain percentage of the FC value, x, (e.g. 20%), the moisture deficit at the end of the day, Dr_n, will be considered 0 and the soil is at field capacity. If θ_{real} is lower than the set threshold, the value of Dr_n will be the higher between $Dr_{real,n}$ and $Dr_{theorical,n}$, so the most limiting value was taken to determine the irrigation timing Eq. (7)

$$\text{If } \theta_{real} > x \cdot FC, \qquad Dr_n = 0$$
$$\text{If } \theta_{real} < x \cdot FC, \qquad Dr_n = \max(Dr_t; Dr_{real}) \tag{7}$$

2.2.4 Irrigation Needs

Theoretical irrigation needs (IN_n) (mm/day) are estimated using crop information, irrigation strategies, climate data and some corrective factors may modify ET_0 (Eq. 8):

$$IN_n = ((ET_{0_n} \cdot K_{c_n} \cdot K_{s_n}) - ER_n) \cdot f_{CE} \cdot f_{Pte} \cdot f_{Rs} \cdot IDC_n \tag{8}$$

where ETc, the crop evapotranspiration, is calculated according to Allen et al. [3], as the product of ET_0 (estimated from weather forecasts) and the crop coefficient, K_C, that depends on the crop and its stage of development. For instance, K_C is conditioned by the ground area shaded by tree crops [8] while for greenhouse crops, it is calculated as a function of the accumulated thermal time (TTA) since emergence [16].

In greenhouse crops, where drip irrigation is used and there is no water input due to rainfall events (ER $= 0$), it can be considered that the water storage in the soil does not change over time. Therefore, irrigation programming is focused on determining when and how much water must be applied to satisfy the crop irrigation needs (ETC) [14, 16]. However, for irrigated outdoor crops, rainwater can partially cover their water needs, so it must be taken into account for irrigation scheduling.

Only a fraction of the rainwater is usable by the crop since part of it is lost by runoff and deep percolation. The amount of water that infiltrates the soil depends on the soil type, and its moisture content at the beginning of the rainfall event, as these parameters determine the soil's capacity to store water. The useful fraction of soil water for plants is known as effective rainfall (ER). There are several methods for ER calculation [3, 12, 48, 53]. Most of these methods (fixed percentage, reliable precipitation and empirical formula) are based on the calculation of effective rainfall on the volume of monthly precipitation. For this reason, a specific procedure to calculate daily effective precipitation based on daily rainfall data recorded in weather stations has been included in the irrigation scheduling process.

To address these limitations, Eq. (9) is proposed to calculate the daily effective rainfall; where Drn-1(mm) is the soil moisture deficit the previous day (Eq. 7), ETc (mm) is calculated from ETo recorded by the nearest agroclimatic station (NAE) and P (mm) represents the volume of gross rainfall recorded at NAE.

$$ER_n = \min(Dr_{n-1} + (ET_{c_n} \cdot K_{s_n}); P_n) \tag{9}$$

For a day n, the storage capacity of rainwater in the soil is calculated by adding to the previous day's capacity (Dr_{n-1}) the increase in storage capacity as a result of crop evapotranspiration ($ET_{Cn})$ calculated using real $ET_{0\,n}$ and gross precipitation (P_n)

recorder by NAE. If the gross precipitation of the day is lower than the soil storage capacity, the total gross precipitation can be stored, so ER is equal to P. In the opposite situation, when ER will reach the value the soil moisture content corresponding field capacity conditions so, ER will be 0. Industrial cultivation conditions are usually achieved in flat fields, so the occurrence of runoff is not considered in the daily basis calculation in Eq. 8. In this case, the water not used by the plant is lost by percolation and evaporation.

The aggregated value of ER_n throughout the irrigation season represents the crop green water use in the water footprint inventory (see Sect. 2.3). To perform the 7-day watering schedule, it is necessary to make ER_n predictions using the same Eq. (9) but in this case, evapotranspiration (ET_{Cn}) is calculated using ET_0 predictions and rainfall forecast.

The coefficients f_{CE}, f_{Pte} and f_{Rs} in equation (8) are correction factors for greenhouse crops as a function of the electrical conductivity of irrigation water, the slope of the greenhouse roof and whitening, respectively [16]. For outdoor crops, the value of these factors is 1.

In water-scarce region, crop water irrigation needs are usually much higher than the water allowances assigned by the water authorities. In these circumstances, the application of deficit irrigation strategies is recommended. They are included in Eq. (8) by the daily irrigation deficit coefficient IDC_n that varies in accordance with irrigation strategies selected:

- Full Irrigation Requirements (FIR): Irrigation events (duration and water depth) are designed to fully satisfy crop irrigation needs with an IDC equal to 1 in Eq. (8).
- Sustained Deficit Irrigation (SDI): Only a fixed fraction of crop irrigation needs is satisfied throughout the irrigation season. In this case, IDC equals this fraction being a constant value during the irrigation season.
- Regulated Deficit Irrigation (RDI): A variable fraction of crop irrigation requirements is applied throughout the irrigation season, according to the crop cycle, avoiding or reducing as much as possible the water stress in its most sensitive crop stages: flowering and fruit set [24]. Thus, IDC depends on the crop development stage being equal to the variable fraction of crop irrigation needs to be satisfied during the irrigation season,

Regardless of the chosen irrigation strategy, the sum of IN_n must be less than the crop water allowance (IWA) for the irrigation campaign. Thus, IDC of the selected strategy is combined with the average fortnightly ETC (historical) to distribute the water allowance according to the crop irrigation needs. Therefore, every fortnight is assigned a fortnightly IWA. The occurrence or not of rainfall together with deviations of ETC from the historical average will condition whether the total amount of water available for each half of the month is used or not.

Any deviation from the established limit will be recorded by the irrigation meter, therefore, at the end of each fortnight, the remaining allocation, if any, will be distributed proportionally to the needs of the remaining months of the irrigation season, readjusting the established limits every 15 days.

2.2.5 Decision to Irrigate

As IWA is distributed fortnightly, it is necessary to fix a daily irrigation limit (IWA_n) to prevent the half-month water use from overpasses that value using Eq. (10):

$$IWA_n = \frac{IWA}{hmd} \tag{10}$$

where hmd is the number of days of the current half-month, with varying values between 13 and 16.

Therefore, the real irrigation requirements of the crop (IR_n), that is, the volume of water to be applied in each irrigation event will be either the theoretical crop irrigation needs (IN_n in Eq. 8) or the daily limit of water availability IWA_n. If IN_n was greater than IWA_n, the irrigation volume to be applied would be IWA_n, otherwise, IRn would be equal to INn. Hence, the added value of IRn throughout the irrigation season will be compared with the water meter records to check the deviations between the expected irrigation volume (irrigation scheduling) and the applied irrigation (water meter) accounting for the blue water use of the crop in the water footprint inventory (see Sect. 2.3).

Once the volume of water to be applied is known, the theoretical irrigation time $t_{t,n}$ (hours) is calculated from Eq. (11) considering the irrigation system efficiency IE (e.g. 0.95 for drip irrigation), the area of the irrigation sector A (ha) and the emitters flow q_e (l/h) and n° of emitters of the irrigation sector n_e [26]

$$t_{t,n} = \frac{IR_n \cdot A \cdot 10^4}{IE \cdot q_e \cdot n_e} \tag{11}$$

The irrigation time cannot exceed the hour limit of the electric rate period assigned to watering, so the recommended irrigation time ($t_{r,n}$) (Eq. 12), is the minimum of $t_{t,n}$ and the number of available hours for irrigation according to the electric rate ($t_{e,n}$):

$$t_{r,n} = \min(t_{t,n}; te, n) \tag{12}$$

The criteria used to program irrigation using the information recorded by the sensors via ITCs is shown in a flow chart in Fig. 3. In this figure, the rounded-edged and thickly drawn boxes indicate that the source of the data is farm-specific information to be entered by the user. The dashed lines indicate that the source of the data is sensors (on-site or remote) that have transmitted the information through the ITCs.

To ensure optimal daily irrigation scheduling, all calculations shown are updated daily with new data on weather forecasts, soil moisture water content measured by sensors and the latest daily rainfall, and ET_0 data recorded at the nearest agroclimatic station. Since the function of IDSS is to provide an optimal irrigation schedule for 7 days, in addition to using the ET_0 predictions to estimate the future irrigation needs of the crop, Eq. (4) is used to make predictions of the moisture depletion level for

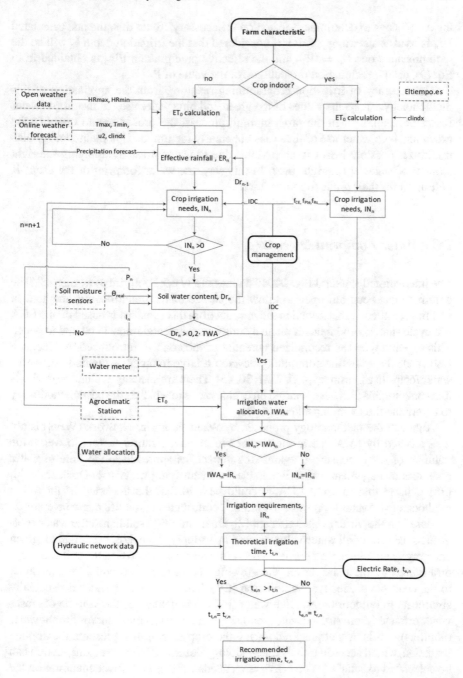

Fig. 3 IDSS flow chart

the next 7 days to determine if irrigation is necessary. To do this, the last calculated Dr_n is used as a starting point; it is considered that the irrigation depth I_n will be the recommended one ($I_n = IR_n$) and the effective precipitation ER_n is obtained from Eq. (9), using precipitation predictions for the value of P_n.

The accuracy of irrigation programming resulting from the application of this methodology affects the values of the green and blue water uses of the crops (water footprint inventory). In the programming process, the procedures to quantify the green and blue water use of the crops are clearly identified. This methodology also minimizes possible losses in crop yields due to water stress by applying the right amount of water at the right time. In this way, the water footprint of the crops is optimized for their production conditions.

2.3   Water Footprint Inventory

The international standard ISO 14046 defines the Water Footprint as a "metric". The definition has been left open to allow the choice of the quantification method. In addition, as already mentioned in the introduction, this standard is based on the LCA life cycle analysis and focuses on analyzing environmental impacts related to water. This together with the recognized potential of the concept introduced by Hoekstra [29] made the scientific community develop a large number of methodologies for water footprint assessment [4, 5, 7, 27, 30, 45]. There are so many evaluation methods that some studies [38] have focused on to analyze as to which is the best methodology to determine the main impact categories.

Although the methodology proposed by Water Footprint Network (WFN) is not well adapted by LCA due to the absence of characterization factors to weigh the volumes of water consumed against the impact, its approach in the field of water resources management makes it suitable for quantifying water use. Hoekstra et al. [28] indicate that to serve LCA, it could be said that the first step of the WFN methodology, "water footprint accounting", contributes to the life cycle inventory.

Based on the virtual water concept [2], Hoekstra [29] establishes the water footprint as the sum of all water volumes used in a supply chain, comprising blue, green and gray water. Green water is defined as rainwater that is stored in the soil and evapotranspired by the plant during the growing period (ET_{green}), which is equivalent to the concept of effective precipitation (ER). This category of water is especially significant in outdoor crops. Blue water (ET_{blue}) is defined as the volume of freshwater extracted from surface water sources (rivers, lakes, ponds) and/or groundwater (aquifers) that is evapotranspired during the crop season. It represents the applied irrigation, which according to the methodology described in the previous section can be calculated to satisfy the crop irrigation needs fully or partially, depending on the crop and the selected irrigation strategy. In short, it is water that is applied to the product and does not return to the environment from which it was initially withdrawn, in other words, it is the water that is "lost" in a particular region [6]. This concept is also known as freshwater consumptive uses. Finally, gray water is an indicator of the pollution impact on water resources and represents the volume of freshwater needed to dilute the pollution to maintain water quality above regulatory water quality standards [29]. There is no scientific consensus at present on a suitable method for quantifying dilution volumes for the assimilation of pollutants, so the estimation of the graywater footprint is subjective [34]. In addition, the graywater footprint is an indicator of environmental impact, so its analysis corresponds to the water footprint impact assessment phase (iii), where the environmental impacts of graywater are more appropriately addressed in other impact categories such as eutrophication or toxicity [40]. In any case, it is not a real volume of water used during production, rather the volume necessary to restore the quality of the water after it has been contaminated throughout the production process, and therefore should not be considered in the inventory of water use [36].

Crop Water Footprint (WF_C) can then be obtained as the sum of the green and blue components, and is normally expressed in m^3/t or in l/kg (Eq. 13):

$$WF_c = WF_{green} + WF_{blue} \qquad (13)$$

The green component of a crop's water footprint (WF_{green}) is calculated as the green component of crop water use (CWU_{green}, m^3/ha) divided by the crop yield, (Y) in t/ha:

$$WF_{green} = \frac{CWU_{green}}{Y} \qquad (14)$$

where CWU_{green} is crop green water use, equivalent to the effective rainfall gathered over the whole production period, which was calculated daily, ERn, according to Eq. (9).

In water-scarce regions, greenhouse roofs are frequently used as rain catchment surfaces. These roofs drain to farm rafts. If a water meter were installed in the drainpipe in the reservoir, green water storage can be estimated and considered as a fraction of the total amount of applied irrigation. Otherwise, in greenhouse crops, CWU_{green} will be considered null.

Unlike other works, the methodology proposed herein uses real-time information recorded or calculated from on-site or remote sensors by ICTs (actual water content in the soil, crop evapotranspiration and daily precipitation) as has been explained in previous sections. For the whole crop season, CWU_{green} is obtained by Eq. (15)

$$CWU_{green} = 10 \cdot \sum_{n=1}^{lds} ER_n \qquad (15)$$

where 10 is the conversion factor of water depth (mm) into water volume per unit surface (m^3/ha), ER_n is the daily calculation of effective precipitation, n = 1 is the first day of the crop season and lds is the first day of the harvest. For permanent crops, $n = 1$ is the last day of the harvest of the previous campaign and lds is the first day of the harvest of the current campaign.

The blue component of the crop water footprint (WF_{blue}, m^3/ha) is calculated as the total volume of irrigation water applied for the whole cropping season (CWA_{blue}, m^3/ha) divided by the crop yield (Y, t/ha) (16).

$$WF_{blue} = \frac{CWA_{blue}}{Y} \tag{16}$$

CWA_{blue} considers the actual total volume of irrigation water applied to achieve the production obtained. It also takes into account the indirect use of water needed for crop production, such as soil preparation, the irrigation dose to compensate for the electrical conductivity of the irrigation water, to reduce frost impact, to remove salt from the soil, to clean the irrigation network, etc. The volume of actual irrigation water applied during the irrigation season will be greater than the theoretical irrigation requirements due to inefficiencies in the irrigation network and soil to provide water to plants when the strategy to fully satisfy irrigation requirements is considered. Over-watering situations are also possible when the applied irrigation is much higher than the irrigation requirements needs plus inefficiencies in the irrigation system and the soil. On the other hand, deficit irrigation strategies may lead to CWA_{blue} below theoretical crop water requirements. Thus, this methodology does not base CWA_{blue} calculation in the theoretical irrigation requirements as many other authors [10, 28]. It is calculated using Eq. (17).

$$CWA_{blue} = \sum_{n=1}^{lds} \left(WA_{g,n} + WA_{p,n} \right) \tag{17}$$

where $WA_{g,n}$ (m^3/ha) is the irrigation water applied to satisfy the crop irrigation needs during the growing period, and $WA_{p,n}$ (m^3/ha) is the amount of water used in different crop management practices. In most studies, the amount of water needed in the different management practices carried out during the crop season to obtain the final products is not considered. $WA_{p,n}$ and $Wa_{g,n}$ are accounted for every day directly from the water meters installed on the farm. Therefore, the amount of blue water used to obtain final crop production (blue water inventory) is obtained by

$$CWA_{blue} = \frac{\sum_{i=1}^{wm} \sum_{n=1}^{lds} \left(I_{n,i} \cdot A_i \cdot 10 \right)}{\sum_{i=1}^{wm} A} \tag{18}$$

where I_n is the daily applied blue water depth (mm) registered by the water meters, i is the sub-index that identifies each water meter, wm represents the total number of water meters installed on the farm and A_i is the area (ha) covered by each one. The denominator of this equation represents the total land area of the farm monitored with irrigation meters.

Considering that CWA_{blue} is applied water, WFA_{blue} (m^3/ha) is the blue water footprint applied, which gives information on the blue water actually used directly and indirectly to produce an agricultural production unit (kg or t), linking it to the crop yield.

It should be remembered that most works on crops' water footprint are carried out on a regional or national scale [4, 38, 39, 45]. At these scales, it is almost impossible to consider the specific cultivation conditions of each farm, unlike the methodology proposed in this chapter. Thus, the calculation of the main components of the WF_C inventory, CWU_{green} and CWU_{blue}, is grounded on the FAO-56 methodology applied through the free software CROPWAT (references and not on real data recorded on-site).

In short, the total amount of water used (blue and green) at field level during the whole crop season (m^3/ha) is the sum of CWU_{green} and CWA_{blue}, Eq. (19).

$$CWU = CWU_{green} + CWA_{blue} \tag{19}$$

Therefore, crop water footprint WF_C is calculated with CWU, Eq. (20).

$$WF_c = \frac{CWU}{Y} \tag{20}$$

Since WF only provides information on how much water has been used to obtain a production unit, it is necessary to know whether that amount is adequate or not. For this purpose, the indicator Relative Irrigation Supply (RIS) [35] is used. It is defined as the quotient between the amount of water applied (recorded in the water meter) and the theoretical irrigation needs during the crop development period, without accounting for the volumes of water used in different crop management practices (indirect water uses) [21, 22, 25]. Thus, RIS values around 1.2 indicate that irrigation has been adequately performed to meet the irrigation needs of the crop taking into account the inefficiencies associated with the irrigation system and the soil. RIS values below and above 1.2 indicate deficit and excess irrigation, respectively. The knowledge of monthly RIS values and for the whole irrigation season allows identifying in which periods the main component of CWA_{blue} (WA_{gn} in Eq. 17) is inadequate and to be able to modify it in the next irrigation campaigns.

$$RIS = \frac{\frac{\sum_{i=1}^{wm} \sum_{n'=1}^{lds} (I_{n',i} \cdot A_i \cdot 10)}{\sum_{i=1}^{wm} A}}{\sum_{n=1}^{lds} 10 \cdot (((ET_{c_n} \cdot K_{s_n}) - ER_n) \cdot f_{CE} \cdot f_{Pte} \cdot f_{Rs})} \tag{21}$$

In this case, n' is the index for the days with irrigation events.

Schematically, Fig. 4 shows the calculation of the WF_C components using the information from the sensors by means of ITCs (dotted line boxes). The scheme includes RIS calculation. The continuous and dotted lines show two different ways of calculating the water footprint.

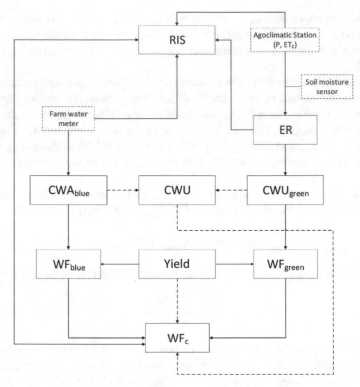

Fig. 4 Crop water footprint (WFc)and Relative Irrigation Supply indicator (RIS) calculation scheme using real-time information via ITCs, where WF$_{green}$ is Green Water Footprint, WF$_{blue}$ is Blue Water Footprint, CWU is Crop Water Use, CWA$_{blue}$ is Blue water applied to the crop, CWU$_{green}$ is crop Green water use and ER is Effective Rainfall

3 Results and Discussion

In the proposed methodology, ICTs are the basic tool used to collect, process the information about water use and manage irrigation in real time. It is used to provide a verifiable and precise inventory of water use based on quality data aimed at improving irrigation water management of two crops cultivated according to the principles of industrial agriculture.

3.1 Irrigation Management Supported by ICT Applications

The irrigation management conditions the values of CWU$_{green}$, CWU$_{blue}$ as well as yield at the end of the crop, when no other problem occurs (e.g. plagues or plant diseases). The application of the developed IDSS to manage precise irrigation systems to irrigate an orange orchard and greenhouse tomatoes for a crop season is described next.

3.1.1 Irrigation Management of an Orange Orchard

The depth of the roots depends on the crop, its state of development, type of soil, presence of tillage sole, among other factors. After an in situ examination of the soil and root system of each crop, sensors were installed at 3 levels in increasing order of depth: surface layer, zone with the highest concentration of roots and the depth at which only a small fraction of the roots reach, where high moisture contents would indicate water leakage by percolation and therefore inefficient use of irrigation water.

In the orange orchard described in Sect. 2.1 dedicated to the "Navelina" variety, the amount of water used by the crop during the 2019/2020 campaign has been analyzed.

As the end of the harvest of this variety is in the middle of January, it has been considered that the campaign of the crop coincides with the year 2019. Therefore, the programming of the irrigation water and the effective rainfall calculated between January 1, 2019 and January 1, 2020 will be used later in the inventory of the water footprint of this crop.

In the year 2019, the gross precipitation (P) recorded by the NAE to the farm was 318 mm.

The data collected by the soil moisture sensors on a daily basis has been used to determine the amount of rainwater available to the crop (CWU_{green}) that can be stored in the soil, as well as the daily soil water balance. Since the data provided by these devices is used in equations involving the water-holding capacity of the soil, it is essential to determine the texture of the soil. In the case of the orange crop, the soil of the study plot was sandy loam. The moisture content at field capacity (FC) is $0.28\ m^3/m^3$ and the permanent wilting point is $0.1\ m^3/m^3$.

Applying the methodology proposed in this work, the irrigation schedule was determined throughout the campaign, comparing it later with the actual irrigation program and consumption of the farm. Weather forecasts from AEMET OpenData and the website eltiempo.es are used to apply the methodology.

In this case, the application during the campaign was conditioned by the water allowance of the campaign ($5000\ m^3/ha$). Two irrigation scenarios were simulated. The first one (Fig. 5) corresponds to the model strategy with an irrigation strategy of full satisfaction of irrigation needs (FIR). The second scenario (Fig. 6) is a simulation of irrigation programming applying a controlled deficit irrigation strategy (RDI). Figure 7 shows the actual management of irrigation on the farm during the study campaign.

In this case, $x = 0.2$ has been defined in Eq. (7) to calculate the soil moisture deficit. The supply of irrigation water is from Fuente Palmera Irrigation district, where irrigation time is fixed between 00:00 h and 08:00 h from Monday to Friday and has no limit on the weekends (24 h).

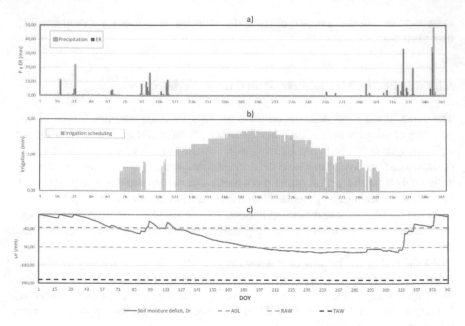

Fig. 5 FIR irrigation strategy simulation (2019). **a** Precipitation distribution, effective rainfall. **b** irrigation scheduling. **c** Evolution of the soil moisture depletion level throughout the crop season with respect to ADL, RAW and TAW

The irrigation strategy in Fig. 5 makes a fortnightly distribution of available water supply according to the historical needs of the crop. Since the water needs of the orange crop are higher than the water allocation for the campaign, the daily irrigation schedule is conditioned by the established water limit (IWA_n). This can be seen in the nearly uniform staggering of irrigation during the season. In any case, even though the strategy is focused on covering the crop's needs, the total irrigation water applied in the campaign is limited by the water allocation. The total irrigation water to be applied following this strategy would be 495 mm, lower than the 500 mm allocation.

To satisfy its water needs, the plant uses water stored in the soil. Water reserves in the soil decrease rapidly during the first months of the campaign due to the low rainfall, bringing forward the start of the irrigation campaign to mid-March. Figure 5a shows that the presence of several precipitation events in spring (approximately from day 90 to day 120 DOY) raised the water content in the soil above the ADL level, so irrigation was not necessary. Later, in the summer months, the water demand of the crop was greater than the water input from irrigation, so the plant took some of the water from the soil to cover part of its evapotranspiration demand.

The application of irrigation was limited by the water allocation (Fig. 5b), so it was not possible to restore the optimal soil moisture level between FC and ADL, increasing the soil moisture depletion level (Dr) until RAW was reached (Fig. 5c). When this threshold was reached, water extraction was more difficult for the plant and the ETC was reduced ($K_s < 1$), so the rate of Dr depletion decreased. In the fall

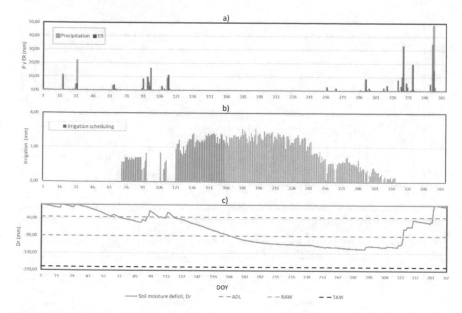

Fig. 6 RDI scheduling simulated (2019). Precipitation distribution, effective rainfall, irrigation scheduling and evolution of the soil moisture depletion level throughout the crop season with respect to the Allowable Depletion Level (ADL), Readily Available Water (RAW) and Total Available Water (TAW)

(approximately 300 DOY) with the arrival of new rains, the Dr decreased increasing the amount of water stored in the soil.

The amount of water that is stored in the soil and usable by the crop (ER) depends on the moisture content of the soil. In Fig. 5a, both P (orange) and ER (green) have been represented. The lower the moisture depletion (Dr) and the higher the precipitation, the lower the ER. Thus, it can be seen, for example, on the 31st day of the year, when the gross precipitation was 22 mm, but only 5 mm is useful for the plant, the rest were lost by percolation and runoff/evaporation. On the other hand, between days 91 and 121, ER equals P because the Dr was high, and the soil had the capacity to store all the rain. The estimated total effective rainfall for 2019 with the use of the FIR strategy was 247 mm.

The application of controlled deficit irrigation strategies (RDI), such as the one shown in the second simulated scenario (Fig. 6), distributed the water allocation according to the crop cycle, making use of an adaptation for irrigation deficit coefficient (IDC) values proposed by Garcia-Tejero [20]. The IDC was used to distribute the available daily water in the daily calculation of the irrigation needs of the crops, avoiding or reducing as much as possible the stress in the periods in which the crop is more sensitive: the flowering and the setting. Unlike the first scenario shown, in this case, the daily irrigation program was not so limited by the allocation of water. Since the water allocation for the season was 500 mm and the gross needs of the

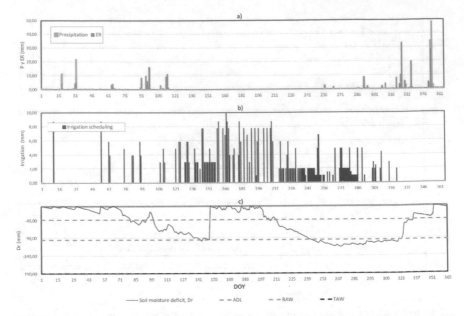

Fig. 7 Real irrigation scheduling (2019). Precipitation distribution, effective rainfall, irrigation scheduling and evolution of the soil moisture depletion level throughout the crop season with respect to the Allowable Depletion Level (ADL), Readily Available Water (RAW) and Total Available Water (TAW

crop were 711 mm (Table 1), the theoretical irrigation needs at critical moments (see Eq. 9) were more restrictive than the IWA_n.

With the RDI strategy, the total water applied for irrigation would be 403 mm, a 20% reduction with respect to the previous strategies as well as greater use of the water stored in the soil. ER takes a value of 266 mm, slightly higher than the previous scenarios.

Table 1 Water Footprint Inventory of the studied crop for the 2019/2020 season

Crop	Orange orchard	Greenhouse tomato
CWU_{green} (m³/ha)	2382	–
CWA_{blue} (m³/ha)	4966	5576
CWU (m³/ha)	7348	5576
IN (m³/ha)	7112	6536
RIS	0.70	0.85
Yield (t/ha)	16.5	109.5
WF_{green} (m³/t)	144	–
WF_{blue} (m³/t)	301	51
WF_C (m³/t)	445	51

The real use of irrigation water (recorded on the water meter) in the 2019 campaign was 497 mm, which is the same as the recommended amount. But the irrigation management done by the farmer (Fig. 7b) differs from the programming obtained with the proposed methodology. Irrigation is less frequent and longer. This affects the evolution of soil moisture, which is significantly different from that predicted with the recommended optimal schedule. During the months of June and July (DOY 153-214), the longer duration and frequency of irrigation events keep the moisture content in the soil at values close to and above FC, indicating that water use is not optimal. With all this, the actual ER calculated for this year was 238 mm.

3.1.2 Irrigation Management of the Greenhouse Tomatoes

The objective of irrigation management in greenhouse crops (e.g. the tomato crop described in Sect. 2.1) is to estimate the needs of the crop to determine how much water and when it should be applied.

In the case of the greenhouse tomato study, there is no pre-established water allocation (Fig. 3). This circumstance avoids the application of irrigation strategies and since the soil's water storage capacity is not considered in this type of crop, irrigation programming is only conditioned by the crop's evapotranspiration.

The crop was developed, in the 2019/2020 campaign, from August 20, 2019 to May 25, 2020. Because water meters were installed after the crop was planted, only water inputs were monitored during the cultivation period, without taking into account previous applications for soil preparation. In this case, the soil moisture sensors have been used only to control the irrigation applied. Figure 8 shows both the real water application during the campaign and the irrigation programming obtained with the procedure proposed in this work.

During the campaign, the irrigation water applied was 558 mm (water meter) and the recommended amount was 654 mm. On this farm, farmers usually reduce water inputs to the crop during the month of the crop harvest. Thus, the main difference occurred in the month of May, when the farmer decides to apply 65% of the water

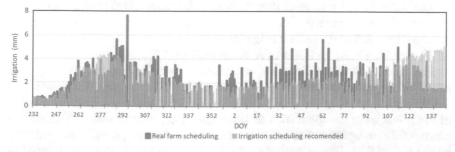

Fig. 8 Real and simulated irrigation management by the model for tomato greenhouse throughout the crop season 2019/2020

needs during the first fortnight and reduced irrigation to 33% of the crop needs during the last 10 days.

3.2 Inventory of Orange and Tomato Water Footprint

The calculation of the WF_C inventory components of the studied crops has been made applying the procedure shown in Fig. 4. The results are presented in Table 1.

The term CWU_{blue} for orange and tomato crops was determined from the records of the water meters installed on each farm. The estimate of CWU_{green} for the crop grown outdoors is based on the estimation of the rainwater amount available for the crop obtained from the measurements of the soil moisture sensors.

The suitability of the irrigation applied to the crops under study during the 2019/2020 season has been evaluated by means of the RIS indicator, in which the irrigation applied for the development of the crop is compared with its theoretical irrigation needs.

The production data of the study campaign were provided by the farm managers: 16.50 t/ha (oranges) and 109.6 t/ha (tomatoes). Although the principles of industrial agriculture were applied to both crops, both farms are classified as organic farms, prioritizing in its production objectives the quality of the product, resulting from a sustainable production process, with respect to the quantity. These same crops produced under non-organic (conventional) agricultural conditions can reach higher values.

The water used in the production of oranges was 7348 m^3/ha, of which 4966 m^3/ha correspond to the water used for irrigation and other cultivation practices necessary to obtain 16.5 t/ha, as shown in Fig. 7. This figure identifies irrigation of important magnitude and frequency between 51 and 200 of the year DOY that led the soil to a level of moisture depletion Dr of 0, being unable to store more water for the crop. A better distribution of irrigation water between DOY 256 and 316 could have prevented the RAW level of water depletion in the soil from being reached, thus avoiding possible production losses.

The water applied was lower than the water allowance (5000 m^3/ha), being also considerably lower than the theoretical irrigation needs of the crop (7112 m^3/ha), as indicated by a RIS of 0.7 (deficit irrigation).

The resulting WF_{orange} is 445 m^3/t, being 68% of WF_{blue}.

The production of tomatoes was not affected by rainwater, so CWU_{green} was zero. With regard to the volume of blue water used (5576 m^3/ha), this is less than the theoretical needs of the crop when it is expected to exceed them as a result of inefficiencies in the application of water and water used in various agricultural practices. These figures indicate deficit irrigation on a campaign scale, as confirmed by the RIS of 0.91. But on analyzing the daily data in Fig. 8, there are periods in which the irrigation water applied to tomatoes is greater than their needs (farmer management), even if it has been applied at the recommended timing. It is important to analyze the daily evolution of CWA_{blue} to improve the efficiency of irrigation

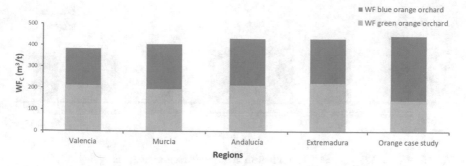

Fig. 9 WF$_{blue}$, WF$_{green}$ and WF$_C$ for the orange grove in different regions of Spain

water use. These deviations from the proposed ideal schedule are usually caused by personnel management issues, as well as other imponderables that may occur during each agricultural season.

The resulting WF$_{orange}$ is 51 m^3/t given the high production of greenhouse crops, much lower than that of the orange crop.

3.3 The Water Footprint of Orange and Tomato Crops in Spain

The Water Footprint of the studied crops for the 2019/2020 season has been compared with values from other studies carried out in several regions of Spain.

Figure 9 shows WF$_{green}$, WF$_{blue}$, and total WF$_{oranges}$ where the values obtained for the studied orange orchard (farm scale) can be compared with the average values of the main orange production areas in Spain [39] between 1996 and 2005 (regional scale).

The highest value of WF$_C$ is obtained in this work as 445 m^3/t (organic farm) but close to the range of average WF$_{oranges}$ values (383 m^3/ha (Valencia) and 431 m^3/ha (Andalucía)). Some of the reasons for these differences may be

- Spatial-temporal scales: In this chapter, WF$_C$ has been at farm scale and for a single campaign, while the other WF$_C$ values correspond to average regional values (climatic and yield data) for a series of 9 years at the regional level (including wet and dry years). The 2019/2020 season studied was a dry season.
- Methodology: WFc was using the FAO-56 methodology that does not take into account neither irrigation efficiency nor indirect water uses during the crop growing season. It is also assumed that crops are irrigated to fully satisfy their irrigation needs. WFU$_{blue}$ usually has lower values than WFA$_{blue}$. This would explain the higher value of WF$_{blue}$ obtained in this work. Likewise, to determine WF$_{green}$, less precise methodologies are used that reduce the rigor and veracity of the data.

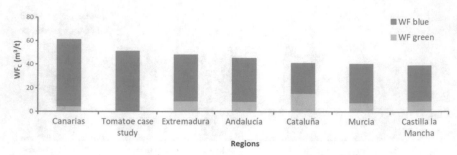

Fig. 10 WF$_{blue}$, WF$_{green}$ and WF$_C$ for tomato in different regions of Spain

In the case of greenhouse tomatoes, Fig. 10 compares the values obtained in this work with those presented by Chapagain and Orr [10] for different areas of Spain in the covered system. Using data from the period 2000–2004, in this paper the authors show crop water use values in different regions in outdoor and greenhouse cultivation systems. In covered systems, they also include tomatoes grown in a mixed system (partly outdoors and partly covered). Therefore, the results presented by these authors also include green water use.

In this case, the highest value of WF$_C$ for tomato in the covered system is in the Canary Islands where WF$_C$ was 61 m^3/t, WFc being 51 m^3/t of the organic greenhouse tomato study in season 2019/2020, followed by Extremadura 48 m^3/t and Andalusia 45 m^3/t.

Likewise, the regional orange WF analyzed above, the differences between the WFc values for tomato obtained in this chapter and those provided by Chapagain and Orr [10] are due to similar reasons.

4 Conclusions

A proper inventory of crops' water footprint at the farm level must be based on rigorous procedures applied to quality data. In this chapter, an appropriate methodology has been developed to obtain the components of the crop water footprint inventory in a verifiable way by cutting-edge technologies.

The developed procedure realizes the optimal programming of the irrigation adapted to the particularities of every farm using ICTs. The required data are recorded by on-site and remote sensors and, sent in real time to the calculation platform via ICTs to be stored in the cloud and accessible for review. It provides information in real time of crop water demands and the evolution of the soil moisture content, to determine the optimal time of irrigation, as well as the quantity of water to be applied so that it can be used efficiently by the crop. The daily and accurate knowledge of the amounts of water used by the crop (green and blue) provides the components of the water footprint inventory of the crop.

The use of an irrigation water use indicator (RIS) allows the evaluation of whether the applied irrigation has been adequate, deficient or in excess. This information is key in the analysis of the components of the crop water footprint. The joint analysis of WF_C and RIS provides both the amount of water needed to obtain a unit of product in certain crop growing conditions and whether or not there suitability of the applied irrigation.

By recording the daily application of water and the evolution of the soil water moisture content, inefficiencies in irrigation management can be detected, so that improvements can be established to contribute to better management of water resources.

The methodology developed has been successfully applied to two organic study farms (orange orchard and greenhouse tomatoes) during an agricultural campaign in southern Spain. Optimal irrigation schedules were obtained for each crop, as well as their corresponding water footprint inventory. These values have been compared with values compiled in the specialized literature on a regional scale in Spain, being values of the same order of magnitude, although more adjusted to the reality of the farms analyzed (water use and production) given the time scale (daily) and space scale (farm) of analysis, the methodology and the sources of data used.

However, the proposed methodology has some limitations in the estimation of effective precipitation on a daily scale, such as not considering the rain characteristics (intensity and duration of each shower). Therefore, the occurrence of runoff or percolation is not considered, as well as the amount of rainwater that could remain on the ground surface after certain rainfall events and infiltrate in the following days, affecting the daily water balance in the soil.

An incipient line of work has emerged from this fact, focusing on increasing the precision of the calculation of rainwater used by crops. A model of runoff generation and soil water movement is being developed that would contribute both to more accurate water footprint inventories and better management of available blue water resources.

References

1. Alcaide Zaragoza C, Fernández García I, González Perea R, Camacho Poyato E, Rodríguez Díaz JA (2019) REUTIVAR: model for precision fertigation scheduling for olive orchards using reclaimed water. Water (Switzerland) 11(12). https://doi.org/10.3390/w11122632
2. Allan JA (1998) Virtual water: a strategic resource global solutions to regional deficits. Gr Water 36:545–546
3. Allen RG, Luis SP, Raes D, Smith M (1998) FAO irrigation and drainage paper no. 56. Crop evapotranspiration (guidelines for computing crop water requirements). Irrig Drain 300(56), 300. https://doi.org/10.1016/j.eja.2010.12.001
4. Bayart JB, Bulle C, Deschênes L, Margni M, Pfister S, Vince F, Koehler A (2010) A framework for assessing off-stream freshwater use in LCA. Int J Life Cycle Assess 15(5):439–453. https://doi.org/10.1007/s11367-010-0172-7

5. Bayart JB, Worbe S, Grimaud J, Aoustin E (2014) The WATER IMPACT INDex: a simplified single-indicator approach for water footprinting. Int J Life Cycle Assess 19(6):1336–1344. https://doi.org/10.1007/s11367-014-0732-3

6. Berger M, Finkbeiner M (2010) Water footprinting: how to address water use in life cycle assessment? Sustainability 2(4):919–944. https://doi.org/10.3390/su2040919

7. Boulay AM, Bare J, Benini L, Berger M, Lathuillière MJ, Manzardo A, Margni M, Motoshita M, Núñez M, Pastor AV, Ridoutt B, Oki T, Worbe S, Pfister S (2016) The WULCA consensus characterization model for water scarcity footprints: assessing impacts of water consumption based on available water remaining (AWARE). Int J Life Cycle Assess 23(2):368–378. https://doi.org/10.1007/s11367-017-1333-8

8. Castel Sánchez JR (2001) Consumo de agua por plantaciones de cítricos en Valencia. Viticultura Enología Profesional Nº 77:27–32. https://dialnet.unirioja.es/servlet/articulo?codigo=159237

9. Castilla N, Hernández J (2005) The plastic greenhouse industry in Spain. Chron Hortic 45(3):15–20

10. Chapagain AK, Orr S (2009) An improved water footprint methodology linking global consumption to local water resources: a case of Spanish tomatoes. J Environ Manage 90(2):1219–1228. https://doi.org/10.1016/j.jenvman.2008.06.006

11. Evett SR, Peters RT, Howell TA (2006) Controlling water use efficiency with irrigation automation: cases from drip and center pivot irrigation of corn and soybean. In: Proceedings of 28th annual southern conservation systems conference, Amarillo TX, September 2015, pp 26–28

12. FAO (2009) CROPWAT 8.0 model

13. FAO (2017) The future of food and agriculture: trends and challenges. In: FAO (ed) The future of food and agriculture: trends and challenges. FAO, Rome, Italy. https://doi.org/10.2307/435 6839

14. Fereres E (1996) Irrigation scheduling of horticultural crops. Acta Hort 449:253–258

15. Fernández García I, Lecina S, Ruiz-Sánchez MC, Vera J, Conejero W, Conesa MR, Domínguez A, Pardo JJ, Léllis BC, Montesinos P (2020) Trends and challenges in irrigation scheduling in the semi-arid area of Spain. Water (Switzerland) 12(3):1–26. https://doi.org/10.3390/w12 030785

16. Fernández MD, Orgaz F, Fereres E, López JC, Céspedes A, Pérez J, Bonachela S, Gallardo M. (2001) Programación del riego de cultivos hortícolas bajo invernadero en el sudeste español. In: Editorial Cajamar

17. Fernández MD, Bonachela S, Orgaz F, Thompson R, López JC, Granados MR, Gallardo M, Fereres E.(2010) Measurement and estimation of plastic greenhouse reference evapotranspiration in a Mediterranean climate. Irrig Sci 28(6), 497–509. https://doi.org/10.1007/s00271-010-0210-z

18. Ferrer M, Viegas M (2014) HUELLA HÍDRICA: La nueva norma internacional ISO 14046:2014 y su implementación. Congreso Nacional Del Medio Ambiente. CONAMA. http://www.conama.org/conama/download/files/conama2014/CT2014/1896712004.pdf

19. Gallardo M, Thompson B, Fernández MD (2013) Water requirements and irrigation management in Mediterranean greenhouses: the case of the southeast coast of Spain. In: Good agricultural practices for greenhouse vegetable crops. Principle for Mediterranean climate areas, November 2015, pp 109–136

20. Garcia-Tejero IF (2010) Deficit irrigation for sustainable citrus cultivation in Guadalquivir river basin

21. García Morillo J, Martín M, Camacho E, Rodríguez Díaz JA, Montesinos P (2015) Toward precision irrigation for intensive strawberry cultivation. Agric Water Manag 151:43–51. https://doi.org/10.1016/j.agwat.2014.09.021

22. García Morillo J, Rodríguez Díaz JA, Camacho E, Montesinos P (2015) Linking water footprint accounting with irrigation management in high value crops. J Clean Prod 87(1):594–602. https://doi.org/10.1016/j.jclepro.2014.09.043

23. Goldstein A, Fink L, Meitin A, Bohadana S, Lutenberg O, Ravid G (2018) Applying machine learning on sensor data for irrigation recommendations: revealing the agronomist's tacit knowledge. Precision Agric 19(3):421–444. https://doi.org/10.1007/s11119-017-9527-4

24. Gonzalez-Altozano P, Castel JR (1999) Regulated deficit irrigation in "Clementina de Nules" citrus trees. I. Yield and fruit quality effects. J Hortic Sci Biotechnol 74(6):706–713. https://www.tandfonline.com/doi/abs/10.1080/14620316.2000.11511256

25. González Perea R, Camacho Poyato E, Montesinos P, García Morillo J, Rodríguez Díaz JA (2016) Influence of spatio temporal scales in crop water footprinting and water use management: Evidences from sugar beet production in Northern Spain. J Clean Prod 139:1485–1495. https://doi.org/10.1016/j.jclepro.2016.09.017

26. González Perea R, Fernández García I, Martin Arroyo M, Rodríguez Díaz JA, Camacho Poyato E, Montesinos P (2017) Multiplatform application for precision irrigation scheduling in strawberries. Agric Water Manag 183:194–201. https://doi.org/10.1016/j.agwat.2016.07.017

27. Helmes RJK, Huijbregts MAJ, Henderson AD, Jolliet O (2012) Spatially explicit fate factors of phosphorous emissions to freshwater at the global scale. Int J Life Cycle Assess 17(5):646–654. https://doi.org/10.1007/s11367-012-0382-2

28. Hoekstra A, Chapagain A, Aldaya M, Mekonnen M (2009) Water footprint manual: state of the art. (Issue January). Water Footprint Network. www.waterfootprint.org

29. Hoekstra AY (2003) Virtual water trade. In: Proceedings of the international expert meeting on virtual water trade, Delft, The Netherlands. Value of water research report series, vol 12. UNESCO-IHE, Delft, The Netherlands

30. Hoekstra AY, Chapagain AK, Aldaya MM, Mekonnen MM (2011) The water footprint assessment manual. Setting the global standard. Soc Environ Account J 31(2). https://doi.org/10.1080/0969160x.2011.593864

31. Hunt CE (2004) Thirsty planet: Strategies for sustainable water management. Zed Books, New York. https://doi.org/https://scholar.google.com/scholar_lookup?title=Thirsty%20Plant-strategies%20for%20Sustainable%20Water%20Management&author=C.E.%20Hunt&publication_year=2004

32. ISO, 14046 (2014) Environmental management—water footprint - principles, requirements and guidelines. In: International organization for standardization, Geneva, Switzerland

33. Jefferies D, Muñoz I, Hodges J, King VJ, Aldaya M, Ercin AE, Milà I Canals L, Hoekstra AY (2012) Water footprint and life cycle assessment as approaches to assess potential impacts of products on water consumption. Key learning points from pilot studies on tea and margarine. J Clean Prod 33:155–166. https://doi.org/10.1016/j.jclepro.2012.04.015

34. Jeswani HK, Azapagic A (2011) Water footprint: Methodologies and a case study for assessing the impacts of water use. J Clean Prod 19(12):1288–1299. https://doi.org/10.1016/j.jclepro.2011.04.003

35. Levine G (1982) Relative water supply: an explanatory variable for irrigation systems. Technical Report No. 6. Cornell University, Ithaca, New York, USA

36. Lovarelli D, Bacenetti J, Fiala M (2016) Water Footprint of crop productions: a review. Sci Total Environ 548–549:236–251. https://doi.org/10.1016/j.scitotenv.2016.01.022

37. MAPA (2019) Encuesta sobre Superficies y Rendimientos de Cultivos, pp 1–178

38. Martínez-Arce A, Chargoy J, Puerto M, Rojas D, Suppen N (2018) Water Footprint (ISO 14046) in Latin America, state of the art and recommendations for assessment and communication. Environments 5(11):114. https://doi.org/10.3390/environments5110114

39. Mekonnen MM, Hoekstra AY (2011) The green, blue and grey water footprint of crops and derived crop products. Hydrol Earth Syst Sci 15(5):1577–1600. https://doi.org/10.5194/hess-15-1577-2011

40. Milà I Canals L, Chenoweth J, Chapagain A, Orr S, Antón A, Clift R (2009) Assessing freshwater use impacts in LCA: Part I—inventory modelling and characterisation factors for the main impact pathways. Int J Life Cycle Assess 14(1):28–42. https://doi.org/10.1007/s11367-008-0030-z

41. Monaghan JM, Daccache A, Vickers LH, Hess TM, Weatherhead EK, Grove IG, Knox JW (2013) More "crop per drop": constraints and opportunities for precision irrigation in European agriculture. J Sci Food Agric 93(5):977–980. https://doi.org/10.1002/jsfa.6051

42. Montesinos P, Camacho E, Campos B, Rodríguez-Díaz JA (2011) Analysis of virtual irrigation water. Application to water resources management in a Mediterranean River Basin. Water Resou Manag 25(6):1635–1651. https://doi.org/10.1007/s11269-010-9765-y

43. Mérida García A, Fernández García I, Camacho Poyato E, Montesinos Barrios P, Rodríguez Díaz JA (2018) Coupling irrigation scheduling with solar energy production in a smart irrigation management system. J Clean Prod 175:670–682. https://doi.org/10.1016/j.jclepro.2017.12.093

44. Odorico PD, Rodriguez-iturbe I (2020) Health of people, health of planet and our responsibility. Health People Health Planet Our Responsib: 149–163. https://doi.org/10.1007/978-3-030-311 25-4

45. Pfister S, Koehler A, Hellweg S (2009) Assessing the environmental impacts of freshwater consumption in LCA. Environ Sci Technol 43(11):4098–4104. https://doi.org/10.1021/es8 02423e

46. Sanz BG (1990) Población mundial y recursos alimenticios. Reis:Revista Española de Investigaciones Sociológicas 49:27–75. https://doi.org/10.2307/40183429

47. Schaap MG (1999) ROSETTA model. J Hydrol 251:163–176

48. Smith M (1992) CROPWAT A computer program for irrigation planning and managment. FAO. https://books.google.it/books?id=p9tB2ht47NAC&pg=PP1&source=kp_read_button& redir_esc=y#v=onepage&q&f=false

49. Smith R, Baillie J, McCarthy A, Raine SR, Baillie CP (2010) Review of precision irrigation technologies and their application. National Centre for …, November

50. UNESCO (2009) The United Nations world water development report 3-water in a changing world (Paris, France). https://doi.org/10.1142/9781848160682_0002

51. UNESCO (2015) The United Nations world water development report 2015: water for a sustainable world. UNESCO Publishing, Paris. https://doi.org/http://www.unesco.org/new/es/natural-sciences/environment/water/wwap/wwdr/2015-water-for-a-sustainable-world/

52. UNESCO (2019) The United Nations world water development report 2019: leaving no one behind. In: UNESCO Digital Library. UNESCO. https://doi.org/10.1037//0033-2909.I26.1.78

53. Villalobos FJ, Mateos L, Orgaz F, Fereres E (2009) Fitotecnia: Bases y tecnologías de la producción agrícola (Ediciones). https://books.google.es/books?id=2kS9V8M03HMC&pg=PA91& hl=es&source=gbs_toc_r&cad=4#v=onepage&q&f=false

54. Vélez Sánchez JE, Intrigliolo DS, Castel Sánches JR (2011) Irrigation scheduling in citrus based on soil and plant water status measuring sensors

55. WEF (2019) Informe de riesgos mundiales 2019. https://doi.org/10.1017/CBO978110741532 4.004

56. Wanjura DF, Upchurch DR, Mahan JR (2004) Establishing differential irrigation levels using temperature-time thresholds. Appl Eng Agric 20(2):201–206

57. Zaragoza CA, García IF, Perea RG, Poyato EC, Díaz JAR (2019) REUTIVAR: model for precision fertigation scheduling for olive orchards using reclaimed Water. Water (Switzerland) 11(12). https://doi.org/10.3390/w11122632

Industrial Water Footprint: Case Study on Textile Industries

P. Senthil Kumar, S. M. Prasanth, S. Harish, and M. Rishikesh

Abstract The concept of water footprint (WF) is an important breakthrough in the evolution of methodologies, approaches, and indicators for measuring freshwater appropriation and assessing the wastewater discharge. Industries have become increasingly aware that they contribute directly and indirectly to water scarcity and pollution, and this constitutes a risk that they have to respond to. Industrial water footprint (IWF) methodology, which concentrates on the industrial production stages, can present a clear graphical view of freshwater consumption and impacts caused by wastewater discharge at both product and environment level. Most of the case studies reported have shown that Industrial Water Footprints (IWF) were caused by grey industrial water footprints. The grey water footprint refers to the volume of freshwater that is required to dilute the toxic pollutant concentration to meet the existing water quality standards. The present study reviews the case study of the textile industry and comprehends the internal water usage information such as the entire plant's water balance, detail of water usage, water yield for water conservation, and recycling measures. Considering a textile and dyeing plant as an example, the water footprints before and after a cleaner production audit were calculated.

Keywords Water footprint network · Cotton textile industry · Textile supply chain · Yarn and fabric manufacturing · Dyeing

Abbreviations

WFP	Total water footprint network (Mt/a)
WFPB	The blue water footprint network (Mt/a)
WFBWE	Amount of blue water evaporation rate (Mt/a)
WFBWI	Amount of blue water incorporation rate (Mt/a)

P. S. Kumar (✉) · S. M. Prasanth · S. Harish · M. Rishikesh
Department of Chemical Engineering, Sri Sivasubramaniya Nadar College of Engineering, Chennai 603110, India
e-mail: senthilchem8582@gmail.com; senthilkumarp@ssn.edu.in

S. S. Muthu (ed.), *Water Footprint*, Environmental Footprints and Eco-design of Products and Processes, https://doi.org/10.1007/978-981-33-4377-1_2

WFLRF   Water expressed in lost return flow rate (Mt/a)
WFPGR   The grey water footprint network (Mt/a)
L[w]   Quantity of pollutant k in the textile industrial network (Mt/a)
$C_s[w]$ The concentration limit of contaminant or pollutant "w" (Mt/a) as specified per standards of pollutant discharge
$C_n[w]$ The concentration of pollutant "w" present in natural water (Mt/a)

1 Introduction

Water is considered to be one of the rudimentary and essential requirements by all living organisms present on earth. It covers a majority of the volume on our planet, occupying about 71% of the whole total surface of Earth. Water is being used as one of the essential components in various industrial processes such as agriculture, construction, thermal power generation, textile manufacturing, paper, and pulp production [3, 31]. Freshwater is a pivotal and highly important natural resource. It is a primary need for living and is an irreplaceable part of human lives and other living organisms [10]. Water is also seen increasingly becoming a global resource, which is due to an increase in the trend of growing international trade in goods and services and its prominence is increasing day by day [9]. When seen at a worldwide scale, it is observed that major water usage occurs in the agricultural sector, but there are also other sectors that add its own significant contribution to the overall water consumption (WWAP 2009) [28]. Figure 1 shows the distribution of water consumption in different sectors. The requirement of water day by day is changing towards a strategy issue at a worldwide, governments of various countries are shifting their primary focus towards water requirement at a sustainable level. By which, the globalization network is starting to use new and complex guidelines and specific and conceptual strategies for the global trade exchange occurring on consumer products by interconnecting various multinational industrial sector by utilizing the data of

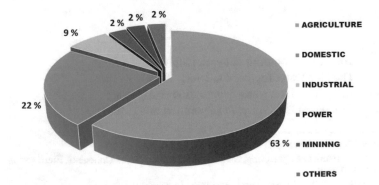

Fig. 1 Water consumption and its distribution network for different sectors

water requirement and consumption pattern and its exchanges. Excluding the regional markets, there is been a significant increase in the global markets for the exchange of water-intensive goods like agricultural products, naturally obtained yarns and fibers, and bioenergy. As a consequence, the use of water-based resources is getting consequently disconnected from consumers. This current trend of globalization of exchange in water-based resources has an enormous consequence on consumers, various governments, and countries and also severely towards nature. Quantity of water is not exchanged except for the minimal amount in the form of containers such as bottles, the water which targeted for specific process or products are exchanged, which is defined as the virtual water, which can affect water consumption pattern in a specified country subsequently.

This is embellished for growing cotton from the field and its processing in textile industries. From agricultural field where the raw material is grown from the fact considering the amount of water required for growth to end product as a fabric for the consumer, cotton as a crop passes through several distinct production steps with different significant impacts on water resources in various forms [8, 21, 34]. Water resource consumption and pollution can be associated with specific activities such as water for agricultural processes, domestic purposes such as bathing, washing, cleaning, industrial purposes cooling of equipment, and for fast processing. Total water that is getting consumed and contaminated is substantially being regarded as the sum of a specified quantity that has independent water demanding activities and contaminating activities. There has been very little awareness given to the impactful fact that, in the end, total water which is getting consumed and the amount of water which is getting polluted will affect directly the people who consume and severely on the structure of the global economy which takes the regard of supplies and exchanges of goods and services network. Until recent years, there has been no thought process or research network in the engineering practices of water network management about how the water is being consumed and various factors by which water is getting polluted along whole production and supply chains. In an all-inclusive image, there has been no to minimal awareness about the industrial organization and its traits and factors which greatly influence the temporal distribution and spatial dissemination of water and its consumption and contamination pattern which can be associated with the end consumer products [8, 20].

2 How Can an Individual Reduce the Water Footprint?

2.1 Direct Footprint Reduction

- By reducing the frequency of chores that requires plenty of water.
- Adopting rainwater harvesting which will replenish the groundwater in the best way.
- Installing water-saving applications inside the household.

- Usage of bucket baths instead of showers.
- Recycled water for cleaning purposes.

2.2 Indirect Footprint Reduction

- Reduction in the consumption of meat products.
- Reduction in the consumption of less water-intensive food and looking forward to season one.

Minimize the usage of harmful chemicals for personal care and cleaning. Human daily activities are causing a momentous impact on the scarcity of available freshwater in the form of an increase in the water pollution level in various forms, and famine to a frightening extent where the future generation will be in a position where the cost of water would be really expensive. These environmental and social issues can only be resolved by contemplating and examining every production step. Hoekstra and Chapagain were among the first few who introduced the concept of WFP which helps in the visualization and identification of hidden water usage in any sector or specific unit and helps in estimating the effects of consumption and trade exchange of water resource. Improvisation and better interpretation of the underlying concepts of water footprint network can form an important basis for the best management and sustainable development of the globe's freshwater resources.

3 Water Footprint Concept

Calibrating the water footprint concept and taking all the required steps to reduce and to keep that level as minimum as possible is extremely important for the environment. The water footprint concept compels us to anticipate indirectly and its embedded water consumption patterns which may occur soon before it's too late. In that sense, water is no longer simply local but tending towards dependent on the global market. There is a desperate requirement of a balance between availability and usage because freshwater resource because water is highly vital to our daily life while the supply of freshwater is limited. As the world population is expanding, so does the gradual need for freshwater is also increasing. Measures and methodologies to decrease the water footprint level, soon becoming the need in the consequence of freshwater water conservation. In an increasingly interconnected yet water-scarce world, this view is particularly relevant.

Calculation of water footprint

$$WFP = WFP_B + WFP_{GR} \tag{1}$$

$$WF_B = WF_{BWE} + WF_{BWI} + WF_{LRF} \tag{2}$$

$$WFP_{GR} = \text{maximum}\left[\frac{L[w]}{C_s[w] - C_n[w]}\right] \tag{3}$$

The "maximum" outside standard brackets means that WFGR which is being quantified by the most or the major critical pollutant contribution that is associated with the largest pollutant-specific WFGR.

Water footprint network can be calculated and expressed in various forms concerning the period of a requirement (per day or per month or per year) which depends on the amount of data required and to what level the information of water footprint is needed, and water footprint is always expressed as the quantity of water in terms of volume per mass of product unit (liter/kg or m^3/ton of products).

To estimate and quantify the water flow network and its network distribution of corporations and products, Hoekstra in the year (2008), introduced the concept of water footprint (WF) as a key performance indicator of the total freshwater volume consumed for the actual process and the amount of water required for controlling the contamination directly or indirectly across a product's whole supply network [19, 20, 29]. Hoekstra also developed a conclusive framework design with accurate equations for quantifying the green, blue, and grey Water footprints of a product and assessing their impact on environmental sustainability.

The Water Footprint is a measure of freshwater appropriation and its underlying importance to a specific product or consumption pattern. These three components are divided into blue, green, and grey WF (water footprint) [5].

- Blue Water Footprint is defined as the amount of surface water and groundwater consumed in the specified process.
- Green Water Footprint refers to the volume of rainwater consumption directly or indirectly by storing it for a specified amount of time.
- Grey Water Footprint refers to the quantity of freshwater required to reduce the concentration of contaminants and maintain the required water quality index as prescribed and agreed by water quality standards globally (Fig. 2).

The estimation and quantification of Water Footprint is a measure of the total amount of water utilized in the production or manufacturing of any product in any industrial unit and by which it can also be scaled up for the whole industrial sectoral

| Volume of rain water consumed | Volume of surface water and groundwater consumed during production processes. | Quantity of fresh water needed to dilute the pollutants and maintain the required water quality |

Fig. 2 Water footprint being classified as Green, Blue, and Grey WF

unit. By the assessment of the water footprint network, the specific area where water conservation methodologies are to be followed is easily identified and can be installed to reduce the overall water footprint of the industry.

The virtual water concept is the trade-in embedded water which was first introduced by Tony Allan which is remarked as the hidden flow of water if food or other commodities are traded from one place to another. Hoekstra has defined the water footprint concept as the volume of freshwater required to produce a specified product. To assess and compare whether the results from the Virtual Water concept and Water Footprint network studies provide insight with complete description regarding the topics of sustainability and its development, efficiency, and equity, we need to focus on the discourse promoted by the Virtual Water concept and Water Footprint network methodologies and towards the underlying ideas that lie at the heart of Virtual Water and Water Footprint concepts.

4   Cotton Textile Industry

The whole textile industrial network is characterized and categorized by the economic activities whose intention is the production of fibers, yarns, fabrics, clothing, and textile products for home and decoration purposes, as well as technical and industrial purposes. Figure 3 shows an overall flow diagram that depicts the processes involved in the production of cotton textile material. However, considering textile fibers as the

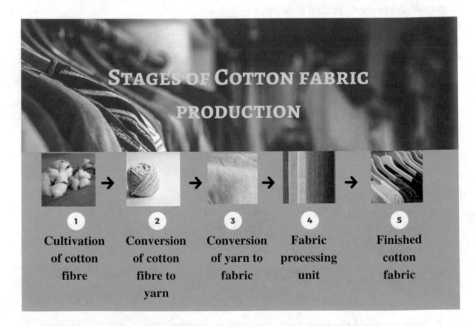

Fig. 3 Different stages of cotton fabric production

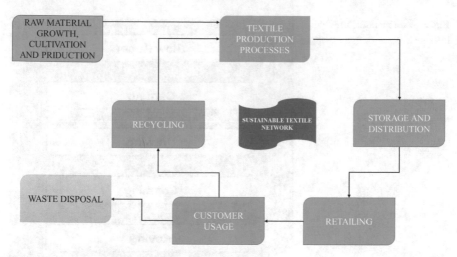

Fig. 4 Sustainable textile supply chain network

basic processing unit of any textile product, textile manufacturing may be identified as the conventional and technical textiles. Among the manufacturing industries present, the Textile industrial network is one of the oldest and most complicated sectors which includes a large number of sub-divisional sectors covering the entire level of the production cycle, from raw materials and intermediate products to the production of final products.

Textile industrial activities present in different subdivisions have their characteristics. Concerning the usage of freshwater resources, the textile industry constitutes one of the most water-consumptive industrial sectors [1, 2, 36], which concludes the fact that there is a need for water footprint to be properly utilized in the textile industrial sector and trace its network. Figure 4 shows the sustainable supply chain network of the textile industry where the water footprint needs to be calibrated for each unit process and operation.

5 Conversion of Fiber to Yarn

The process of converting cotton fiber into a yarn involves a series of steps that help to clean, remove short fiber, alignment of fiber, ultimately spin the yarn. Various steps involved in the processing of fiber to yarn are mentioned in the Fig. 5. Most of the operations performed in this section are dry processes, in which the usage of water is minimal or null. These processes mentioned below varies from industry to industry and also with the quality of fiber cultivated.

Blow room operation consists of blowing of air preferably at high temperature than the cotton fiber collected from the field, to remove the moisture content present in the fiber. Then the fiber is conveyed to the gin stands, in which they are passed through

Fig. 5 Yarn manufacture process

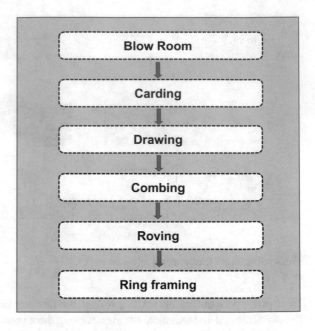

a rotating circular saw which separates cotton seeds from the lint, this process is known as ginning. The carding process further helps in alignment of the cotton fiber into parallel strips using the carding machine and finally results in the production of continuous single-stranded fiber known as a sliver. Then in the drawing process, the fibers are blended, straightened, and the number of fibers in sliver is reduced to achieve the desired linear density of the fiber for the spinning process. The drawing process also improves the uniformity and homogeneity of the fibers. Combing is an optional process in yarn manufacturing as it depends upon the financial background of the individual textile mill. The combing process additionally removes short fibers known as nails and makes the fiber smoother, finer, and uniform than that obtained from the carding process. But combing is costlier than the carding process as it involves additional processing steps and produces more wastes. Roving is the penultimate step before the spinning process. It condenses the sliver to an even finer strand known as roving. The roving is of a few millimeters thickness and consists of knots in between to keep the strands together.

6 Production of Fabrics from Yarn

The process of conversion of yarn to fabric can be accomplished by using two different methods, named weaving and knitting processes. Both these methods involve unique interlacing patterns via which yarn is converted to fabric. This section describes the basic mechanisms involved in the production of fabrics.

7 Weaving Fabric Manufacture

Woven fabrics are produced by the process of weaving using machinery named looms. In this process, two different types of yarns known as the warp yarn (placed parallel to the loom) and the weft yarn (placed perpendicular to the loom) are interlaced together to produce a fabric. Before the actual weaving process, the setup has to be prepared that involves warping, sizing, draw-in, and size-in processes. Warping or beaming is the process of combining yarns from different cones together to form a sheet where racks of bobbins are set up to hold the thread while it is rolled onto the warp bar of the loom. Warping is done to preserve the yarn elongation and maintain a uniform level. Sizing is a process of adding chemicals to reduce the friction between the yarns that generally result in yarn rupturing. Some of the common chemicals used are polyvinyl alcohol (PVA), starch, carboxymethyl cellulose (CMC), etc. Once the setup is completed the interlacing of fabric takes place via shedding, picking, and beating. Factors such as quality of yarn, looming time, and type of machinery used to play a major role in producing a good quality fabric.

8 Knitting Fabric Manufacture

Knitting is the process where the yarn is converted into fabric by interloping the loops present in the yarn. When a loop is drawn through another, loops are formed either in the vertical (warp knitting) or horizontal (weft knitting) direction. The intermeshing of loops in the vertical direction is called wales while the same in the horizontal direction is called a course. In the warp knitting, there are many pieces of yarn and there are vertical chains, zigzagged together by crossing the cotton yarn. While in the weft knitting very similar in method to hand knitting with stitches all connected horizontally. There are two different types of machines used for knitting the yarn to fabric and they are named as flatbed and circular knitting machines. The usage of these machinery varies with respect to production and patterning.

9 Fabric Processing Unit

The fabric in its loom state not only contains impurities, including non-uniform warp size it requires further special treatment to develop into full textile potential. Also, to make the fabric more attractive, coloring agents, which are known as dyes are added after the pretreatment processes. Moreover, final alteration is made in the finishing process to make the fabric into a more commercial product. These processes utilize high amounts of water for rinsing and washing the fabrics in the intermediate steps between each operation. The overall process that involves pre-treatments, dyeing, and finishing is called the fabric or wet processing step.

10   Pretreatment Techniques

Several pretreatment techniques are performed in the textile industry and its requirement varies with the fabric production method. The basic need for pretreatment
is to remove impurities that adhere to the fabric material and also to improve its
surface characteristics such as uniformity and reactivity with coloring agents (dyes).
A detailed description of the pretreatments involved are discussed in the following
section.

10.1   Singeing

Singeing is the process that is designed to burn off the surface fibers from the fabric
to produce fine smoothness.

10.2   Desizing

Desizing is the process that involves the removal of chemicals that were added in the
sizing operation. These chemicals inhibit the dyeing and printing of fabrics as these
chemicals might react with the dyestuff to result in unwanted by-products. Either
hydrolytic or oxidative processes are employed to remove these chemicals from
the fabric material. The hydrolytic method includes rot, alkali, acid, and enzymatic
stepping, while the oxidative method includes bromide, chloride, and ammonium
persulphate desizing. The acid stepping process uses inorganic acid such as dilute
hydrochloric or sulphuric acid for hydrolyses. However, this method of desizing
affects the fiber configuration in cotton fiber. Enzymatic hydrolysis is also empha-
sized in the desizing operation which utilizes enzymes attained from vegetables and
animals. Enzymes are target specific species that act efficiently in the removal of
sizing agents, thus the number of enzymes required will be very less. The problem
with enzymatic hydrolysis is that it loses its activity due to poisoning by foreign
substances. The oxidative method uses oxidizing agents for the removal of starch
from fabric materials. It is highly effective in degrading the sizing agents, however,
when used in the excess amount it destroys the structure of the fiber. After the
desizing operation, the surface of the fabric is washed thoroughly to separate the
traces of sized materials. After washing, the material is subjected to thermal drying
in steam producing apparatus.

10.3 Scouring

Scouring is a chemical washing process carried out on the cotton fabric to remove natural wax and nonfibrous impurities. Naturally occurring impurities such as noncellulosic components, wax, as well as added impurities such as dirt that are present in the fiber and fabrics, are removed by this process. Scouring is performed in the iron vessels called kiers where the fabric is enclosed and boiled in an alkali solution of sodium hydroxide which forms a soap with free fatty acids. An excess amount of water is then used to wash off these impurities from the surface of the fabric. This method improves the absorptive capacity of the fabric material which enhances the dyeing process that comes in the consecutive steps. Also, this process whitens the color of the material without damage to the fiber structure. The effluent water from this process consists of detergents, pectin, oil, and solvents from various other processes.

10.4 Bleaching

Madhav et al. [23]. Bleaching is generally employed in both batch and continuous modes of operation. The degree of bleaching is determined by the required level of whiteness and absorbency.

10.5 Mercerising

Mercerization makes the fabric more lustrous and its tensile strength and dye uptake capacity. It also reduces the shrinkage of fabric and gives a silk-like texture. This process is carried out by submerging the cotton fabric in a medium of sodium hydroxide (15–25%). It results in the phenomenon of swelling due to the absorption of sodium hydroxide by the cotton fiber. After swelling, the basic solution used is washed off thoroughly either by using an acid solution or water itself. During this process, the fabric is subjected to vertical deformation which is prevented by applying tension on both ends of the fabric material. Sodium hydroxide used in this process is recovered in most of the industries by membrane assisted technologies.

10.6 Dyeing

Dyeing is the final process where the absorbent fiber responds readily to the coloration of the final textile product. Colorants are added in the industry to impart color to the cotton fabric material. Dyes are colorants that have a natural affinity towards the fabric material and remain permanent within it. These colorants diffuse into the molecular structure of the fabric and produce color to the fabric material. The type of interaction between the dye and the cotton fiber determines the color sustainability of the material. There are different classes of dyes present in the world both natural and synthetic, with over 1000 types of dyes currently used in industries. The most common classes of dyes used in the cotton textile industry are represented in Table 1.

These dyes differ from each other due to variation in their characteristics and choosing a specific dye for a process requires the following considerations: based on its availability, cost, application, and effects it leaves in the environment. Other types of colorants used are known as pigments. These pigments do not have a natural affinity towards cotton fiber and therefore additional chemicals must be added to glue them to the cotton fabric. These additional chemicals are called binders as they act as a bridge between the pigment and fiber molecule. Along with dyes, auxiliary chemicals such as salts, surfactants, heavy metals, etc., are added to the dye together with the cotton fiber. Water is used to carry these colorants in the form of steam to settle in the baths. After the dyeing process is complete, the water present in the water

Table 1 Detailed insight about different classes of dyes used in the cotton fabric industry

S. no		Types of dye used in the cotton textile industry			
		Reactive	Direct	Vat	Sulfur
1	Mechanism	Reacts directly with the chemical structure of cellulose	Does not react directly	Forms of an aqueous dispersion	In the presence of a reducing agent, the dye particles disintegrate
2	Nature of bond	Covalent	Hydrogen bonding/dipole interactions	Covalent	Hydrogen bonding/dipole interactions
3	Solubility	Highly water-soluble	Highly water-soluble	Requires chemicals to attain high solubility	Requires chemicals to attain high solubility
4	Brightness	Bright	Fairly bright	Dull compared to reactive and direct dyes	Wide shade range but dull colors
5	Fastness	Does not fasten upon washing	Poor washing fastness	Does not fasten upon washing	Dark shades don't fasten on washing
6	Auxiliary chemicals	A large number of salts	Less compared to reactive dyes	Requires reducing agents to form a reactive mixture	Requires reducing agents to form a reactive mixture
7	Percentage of unfixed dye	20–50	5–20	5–20	30–40

bath consisting of numerous amounts of colored impurities and organic salt content is released out. Nearly 30% of dye concentration does not get fixed to the surface of the fabric, thereby contributing to the contamination of water. This also leads to an increase in the water quality parameters and this is an indication of water pollution when it's released into the environment as the same. The amount of water used in the dyeing depends on the type of dye used and mainly the machinery employed to accomplish the process. Some of the major types of machinery used for dyeing fabrics in the industry include jet, jigger, pad, beam, and solvent dyeing machines. Each method has its advantages and disadvantages and its use in the industry depends on the factors similar to those listed for the dye selection.

10.7 Printing

This process is employed to produce selective coloration of fabric materials consisting of different designs. Unlike dyeing, printing involves the usage of dyes in semi-solid form like that of a paste to avoid smudging of color out of its range. It is also referred by the term localized dyeing. Though the number of colorants used might be less, it produces a similar proportion of BOD contamination in the wastewater compared to dyeing. After the coloration, the fabric material is washed to remove the excess freely.

10.8 Finishing

This operation uses a wide range of mechanical, chemical, and enzymatic methods used for the finishing of fabrics. The finishing operation is performed for improving the physical appearance, softening, and waterproofing of the fabric. Also, the fabrics are disinfected by adding disinfectants named biocides to mitigate the microbial growth. Most of these finishing stages pollute the water as they utilize hazardous chemicals such as formaldehyde. However, due to the excellent activity of formaldehyde towards improvising the quality of fabrics in producing stiffness, smoothness, and dimensional stability [6]. So, the search for an alternate, instead of using formaldehyde, is still going on.

11 Water Footprint Network of Textile Industry

Cotton and cotton textile industries are central to the economic growth of both developed and developing countries. Almost 50% of the textile industry depends upon the cotton raw material for survival. Though various artificial fiber such as nylon,

Fig. 6 Global average water
footprint spread for cotton
cultivation [24]

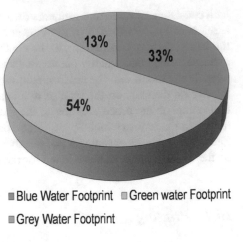

■ Blue Water Footprint   ■ Green water Footprint
■ Grey Water Footprint

etc., have been developed, cotton remains the primary raw material for fabric development in many parts of the world. Water footprint, in the case of manufacture of cotton fabrics, can be divided into two categories, the water footprint for growing and harvesting of cotton (agricultural process) and water footprint for the conversion of cotton fiber to fabrics (Industrial process) [35]. These multiple stages of cotton fabric production may take place at a single place or different locations because agricultural processing of cotton may not be viable in some parts of the country. Cotton seeds are grown in Mediterranean regions and other warm places across the world where freshwater is available in raft supply. Like any other water footprint network, cotton's footprint also consists of three classifications of water namely green, blue, and grey. Each of the processes in the manufacturing of cotton fabrics that is both the industrial and agricultural processes accounts for separate green, blue, and grey water footprint. Figure 6 represents the Global average water footprint used for cotton production.

12   Agricultural Processes

The large area under cultivation of cotton crops makes it the second most significant crop cultivated in terms of land use after soya bean [14]. Cotton is primarily produced for its fiber, which is used as a textile raw material. In 2013–14, cotton was cultivated approximately 32,429,000 ha (ICAC, 2014), equivalent to about 2.3% of the world's arable land (almost 1.4 billion ha). The majority of the cotton produced, that is almost 80% of all cotton is produced in six countries. China is the world's leading producer, followed by India, the United States, Pakistan, Brazil, and Uzbekistan. China and India account for slightly more than half of world cotton production, while the United States, Pakistan, Brazil, and Uzbekistan accounted for an additional 29% [15, 33].

12.1 Blue and Green Water Footprint in Cotton Cultivation

Water footprint in the agricultural processing of cotton crops can be ascribed mainly to water used for the growth of cotton crops, which mainly depends upon the water obtained from rainfall and irrigation water supplied. The evaporation of infiltrated rainwater attributes to the green water footprint, whereas the surface or groundwater extracted for irrigation constitutes the blue water footprint. The ratio of green to blue water footprint in agricultural activities may vary significantly based on the climatic and growth conditions of a particular region. Tropical places may receive higher annual rainfall which may increase the green water consumption comparatively. Places with limited rainfall periods annually may depend more upon the irrigation water for the growth of crops.

The total Blue water footprint of the agriculture process is given by

$$WF_{blue} = Bluewaterevaporation + Bluewaterincorporated$$
$$+ lostreturnflow \left(\frac{volume}{time}\right) \tag{4}$$

Here the lost return flow refers to the water lost to other catchments as runoff and which is not available for reuse within the same period [22]. The total green water footprint of the agricultural process is given by,

$$WF_{green} = GreenwaterEvaporation$$
$$+Greenwaterincorporation\left(\frac{volume}{time}\right) \tag{5}$$

Here the green water evaporation refers to the total rainwater lost by evapotranspiration and green water incorporation refers to the water incorporated into the harvested crop [11].

Irrigation is the major contributor to the blue water footprint in the case of the agricultural process. Around the world, approximately 6% of total irrigation water is used for its cultivation. In countries like India, irrigation attributes to 80% of total water usage in the country [7]. Also, the old and less efficient irrigation system can increase the amount of blue water used for cotton cultivation. Hence, improvements and innovation in existing irrigation systems to minimize water usage without compromising on the yield of the crops can tremendously conserve the blue water footprint.

Generalization of the amount of green and blue water footprint of the cotton crops is a byzantine task. Though various data regarding the average annual rainfall at a certain place, amount of irrigation water supplied per hectare of the field, etc., are available on a global standard, these data are anachronistic and may not even pertain to the current climatic conditions. But these data can serve as an early prediction or an initial guess to further calculate the actual and pertinent water footprint measures. Another important term in the estimation of water footprint is "Virtual water". It is defined as the quantity of water used to produce a commodity [4]. It is also known as the embedded water which is analogous to the embodied energy concept mentioned in [18]. Virtual water comprises all domestic water usage and international water usage as well. Foreign water usage comes into the picture when the import and export of cotton crops are considered. Various data regarding the generalization of calculation of water footprint has been systematically computed in [21]. Research in [21] takes into account of top 15 cotton-producing countries across the world and gives a comprehensive report of the virtual green and virtual blue water used for cotton growth in these countries. CROPWAT model is utilized to estimate the effective rainwater and irrigation water used in these countries. The virtual green water footprint is calculated as the ratio of effective rainfall to that of crop yield and the virtual blue water footprint of a crop is calculated as the ratio of effective water used for irrigation to that of the yield of that crop [21].

12.2   Grey Water Footprint in Cotton Cultivation

Apart from the blue and green water footprint, the leaching of fertilizers and pesticides adds to the account of grey water footprint. Though the grey water footprint is mainly associated with the industrial processes, this concept indicates the level of pollution and contamination in the cotton crop growth and harvesting process. Grey water in agricultural processes is defined as the quantity of water to assimilate or dilute this leached water due to fertilizers so that the final concentration of contaminants is below the permissible level. On a global level, the cotton fiber results in extensive usage of insecticides and herbicides accounting for almost 16 and 6.8% of all insecticides and herbicides used globally [26]. The primary pollutants present in leached water are nitrates and phosphates that are the primal components of fertilizers and pesticides. These leached chemicals present in the soil may penetrate the soil and debase the ground and surface water. Furthermore, nitrate and phosphate in the surface water can act as the nutrient source for many micro and macro algae which further reduces the dissolved oxygen present in the water. Among the other pollutants, nitrogen is more susceptible to leaching. It is not readily absorbed by the soil and hence results in movement through the layers of soil. Comparatively, phosphorus has lower mobility than nitrogen, and hence leaching of phosphorus is not a major problem, but they react with other minerals present in the soil and form insoluble compounds. Hence, the estimation of the grey water footprint of an agricultural process may help in mitigating the harm caused to the surface and groundwater. Grey water footprint for a growing cotton crop can be calculated as

$$WF_{grey} = \frac{L}{C_{max} - C_{natural}} \left(\frac{volume}{time} \right) \tag{6}$$

where L is the pollutant load reaching water bodies (in mass per time), $C_{natural}$ and C_{max} are the concentration of pollutants (in mass per volume) naturally found in the surface water and maximum permissible pollutant concentration in the surface water, respectively. In the case of agricultural activities, the major pollutants are leached fertilizers.

12.3 Effect of Agricultural Practices on the Water Footprint of Cotton Production

A thorough examination of the water footprint of raw materials and their processing is very much important for minimizing water usage and making the overall process more sustainable. It was identified that more than 60% of the cotton agriculture's blue water footprint lies in India [7]. It was found that out of all the cotton cultivated lands in India, 39% of it belongs to a specific region called the Indus river basin which is a major hotspot for water scarcity problems. Hence, the identification of such hotspots and problems arising from these hotspots are very much important for a sustainable water footprint. The water footprint associated with cotton agriculture varies with varying cultivating practices. Two major agricultural practices include conventional cultivation and organic farming.

Conventional farming is a common agricultural practice that is employed extensively in various parts of the world. Approximately, 90% of all cotton is grown conventionally. Conventional farming can be rain-dependent or irrigated. It employs a combination of synthetic agrochemicals for pest control and fertilizers for enhanced or quicker growth. The restriction placed on the number of chemicals used is languid. Conventional cotton uses about 16% of the world's insecticides and 7% of the world's pesticides [7]. Variation within conventional farming can be observed from farm to farm and from different geographic locations as there are no guiding principles involved. Spraying of chemicals is often done on an extensive scale on all plants, timed according to a prescriptive schedule. Although many conventional farmers apply good agricultural practices and use fewer chemicals, they are not certified or verified as being more sustainable.

Organic farming includes various techniques like crop rotation, biological pest control, composts, etc., to make the process of agriculture of cotton eco-friendlier and more sustainable. The chemical usage of fertilizers and pesticides is minimal and is used only when these chemicals are derived from a natural source. Though eco-friendly, this practice only accounts for 10–15% of the total cotton production. This is because organic farming is very time consuming and also produces low product yield. These two different practices give rise to a very different water footprint. Conventional farming uses a profuse amount of fertilizers which extensively increases the grey water footprint for such practices. Whereas in the case of organic farming more importance is given to green and blue water footprint since there is minimal usage of chemicals that harm the soil and surface waters. It was shown in the study [16], that although the yield of conventional cotton per year in India, was slightly higher than that of the organic cotton, conventional farming produces 5.5 times the grey water footprint produced by the organic farming. Hence, on comparing these two farming practices, a shift towards organic farming seems to be a more sustainable option as it drastically minimizes the grey water footprint and also produces comparable yields of cotton. But still, the research on the effects of organic fertilizers on the water resources is limited and should be actively pursued to shift our agricultural practices.

13   Industrial Processes

Water is one of the essential resources that is utilized in a large amount in the cotton textile industry. Most of the water is used to rinse and wash the fabric produced by weaving or knitting processes. Also, water is used in cooling operations in the wet processing unit for drying the fabric after subsequent washing operations. As shown in Fig. 7, around 72% of the water is consumed in the processing stage while a significant portion is utilized for cooling, steam generation, and other miscellaneous operations.

The textile industry consumes water either from surface resources such as ponds, lakes, rivers, or groundwater resources. This form of water contributes to the blue water footprint. The significance of green water footprint in the industrial processing of cotton is very less since reliance on rainwater is not feasible due to the sporadic nature of rain around the world. Therefore, in the cotton textile industry blue water footprint is higher compared to the green water footprint. According to the IPCC

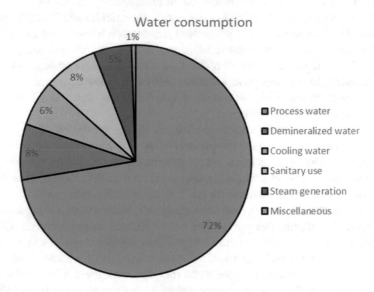

Fig. 7 Percentage split-up in water consumption of the textile industry [25]

textile BAT Reference (BREF), the specific water consumption for a cotton textile industry lies in the range of 10–645 L/kg of the final product [27]. Also, huge quantities of chemicals are added to the wet processing unit of the textile industry. The specific consumption of dye and other chemicals used in the processes are quantified to be 10–50 and 58–670 kg/kg of final product produced [13]. These chemicals are washed off from the fabrics by the action of water. Therefore, the effluent water gets contaminated with these chemicals and it contributes to the grey water footprint. Industrial processing forms the core of textile industries and is the key consumers of blue and grey water footprints.

13.1 Water Utility: Process Operation

Based on the consumption of water in the process operations, the processes in the textile industry are classified into dry, intermediate, and wet processing units [12]. Dry processes are the operations that consume a minimal amount of water and produce no significant wastes. Intermediate processes also consume water but the amount of waste disposal is not significant compared to the wet processes. Wet processes consume water and most of the water consumed is discharged as waste. Hence, the majority of the water footprint in the case of industrial cotton processing can be traced to the intermediate and wet processing of cotton fabrics [32].

13.2 Dry Processing Unit

Processes that consumes very less or negligible water footprint in the textile industry are present in this category. These include spinning, weaving, and knitting process which involves the production of yarn and fabrics from the cotton fiber. Also, a significant amount of water is consumed only in the process of blending of fibers of different varieties to improve the quality of yarn produced. But this amount is also negligible when compared to the water footprint of the intermediate and wet processes involved in fabric manufacturing. Therefore, the operations that are termed as dry processing are not considered for the water footprint calculation.

13.3 Intermediate Processing Unit

The processes that consume a considerable amount of water in carrying out operations and produce grey water less than the wet processing stage falls in this category. Sizing is one such operation where its specific water consumption is in the range of 0.5–8.2 L/kg of product. Water in this process is used to mix with sizing agents and then is applied to the yarn. During the operation, a certain amount of sizing liquid is wasted in the process. Though its volume is not considerably large (i.e., less grey water volume), however, it has high levels of COD, BOD, and SS which pollutes the environment. Another important operation that comes in this category is the printing process. The specific water consumption of this process lies in the range of 8–16 L/kg of the product. In this process, water is used to wash off the colorants added in specific locations, according to the design pattern. The amount of water used for printing is very less compared to that of dyeing. However, the colorants used in this process contaminates water with high levels of COD and TSS.

13.4 Wet Processing Unit

As the name suggests the wet processing unit utilizes large amounts of water, majorly for rinsing the fabric after chemical treatment. Also, water consumption varies widely with the equipment used to carry out the process. For example, the water requirement for a batch processing unit is proportional to the bath ratio and other characteristics such as mixing, agitation, and the turnover rate of fabric [30]. In general, the wet processing unit comprises pretreatment, dyeing, and, finishing operations that are specific to the type of fabric used. Therefore, the basis of the classification of the water footprint of the cotton textile industry is built on the different methods of fabric production. The basic steps involved in the wet processing stage of woven and knitted fabric is shown in Fig. 9. The major footprints: blue and grey water footprints for each of these fabrics are compared and discussed in detail in the following section.

14 Blue Water Footprint in Textile Industrial Processes

The blue water footprint of the fabric processing unit is described in Fig. 8. It is evident from the figure that the dyeing operation utilizes a tremendous amount of water compared to the other processes. The desizing and bleaching operations consume water in the range of 2.5–25 L/kg fabric. Scouring and Mercerization consume water in the range of 20–45 and 17–32 L/kg fabric. Dyeing requires water in the range of 10–300 L/kg of the product. The reason for the range to be larger is because the water requirement varies according to the chemical used for dyeing and the machinery assigned to perform the dyeing operation. The water requirement of some of the dyeing equipment is listed in Table 2.

The specific water consumption for dyeing of the woven fabric process is found to be in the range of 130–150 L/kg of dyed fabric. While for the knitted fabric process, it is found to be in the range of 110–130 L/kg of dyed fabric. The woven fabric has a higher water footprint than the knitted fabric due to additional pretreatment operations involved in the wet processing unit as described in Fig. 9. This high-water consumption will reflect in the environment in the form of grey water which is released as effluent from each of the individual processes.

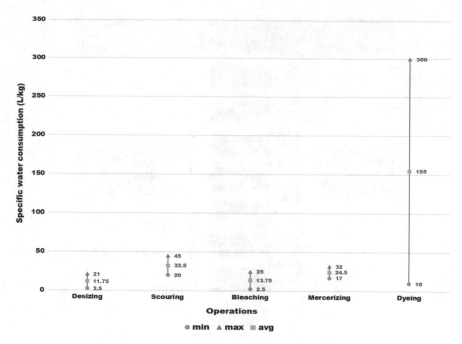

Fig. 8 Represents the specific water consumption (L/kg) of each process in the wet processing unit

Table 2 Major dyeing machinery and their corresponding liquor ratios used for cotton fabric dyeing

S.No	Equipment type	Material to Liquor Ratio (MLR)
1	Jet	1:7–1:15
2	Jig	1:5
3	Beam	1:10
4	Continuous	1:1
5	Beck	1:17

14.1 Grey Water Footprint in Textile Industrial Processes

The effluent water produced in the wet processing unit is quantified based on specific water consumption by multiplying the total kilograms of the finished product by the water consumption of the corresponding processes. This effluent water contributes to the grey water footprint of the wet processing unit. The average grey water footprint produced in the wet processing unit for the production of woven and knitted fabric is 16,000 and 3,300 L/kg product as shown in Fig. 9. Higher volumes of grey water production in woven fabrics is attributed to the additional pretreatment requirements as compared to the knitted fabrics. The grey water consists of chemicals used in the wet processing unit and these chemicals are highly harmful to the environment. Along with these chemicals, the unfixed dyes will also be present and their percentage

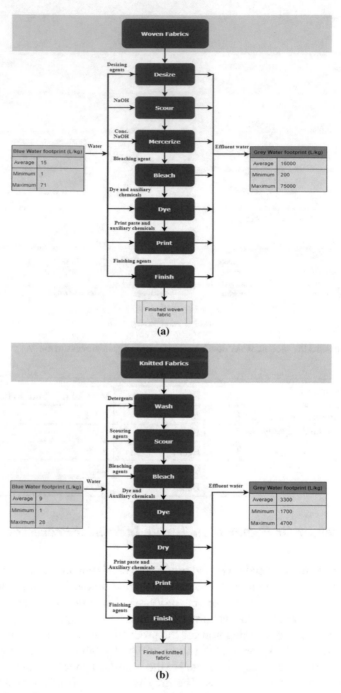

Fig. 9 A schematic representation of processes involves in woven (**a**) and knit (**b**) fabrics in the wet processing unit along with its water footprint [16]

has been given in Table 1. The unfixed dye is removed from the fabric by performing several wash cycles and is aided by adding detergents.

These chemicals affect the quality of water as they are presently more than the permissible limit. The TDS, TSS, BOD, and COD levels are found to be higher for woven fabric than knitted fabric. Also, the presence of dye imparts color to the grey water which inhibits the penetration of the sun's rays to the living organism, when it is directly disposed of in a water body. This decreases the photosynthesis of aquatic organisms and that leads to a decrease in the dissolved oxygen (DO) level of the corresponding water body. Proper treatment techniques have to be performed to reduce these contaminants as they are considered to be hazardous to the environment.

14.2 Water Utility: Other Operations

There are several places in the industry other than the process operation, where water is required. In the textile industry, certain operations require water with at most quality, i.e., demineralized water. Water from obtaining from surface resources (which is blue water) contains a high amount of salts, which results in the hardness of water (hard water). To remove the hardness from the water, the reverse osmosis (RO) mechanism is employed in most of the industry. About 8% of the overall water consumption is contributed by this RO process to produce water with minimal hardness (soft water). Another important operation carried out in the industry that consumes water is cooling. About 6% of the water consumption is attributed to the cooling operation. Cooling is done to maintain the temperature and humidity in the textile industry. The textile industry might run all year round and the humidity varies every day. Depending on the humidity of air present in the textile industry, the absorption or release of moisture from the cotton fiber can be determined. This cycle if not properly maintained will affect the characteristics of the cotton textile produced in the industry. The specific water consumption of cooling operation lies in the range of 8–12 L/kg product. To maintain the temperature, i.e., to heat the water baths in the wet processing unit, steam is used. Steam is produced by heating water in a boiler to a temperature greater than 100 °C. About 5% of the overall water consumption is attributed to this operation. The water used to produce steam should have lesser salts because with time these salts settle at the bottom of the boiler leading to inefficient heating. This phenomenon is called scaling and in extreme case scaling may even lead to boiler explosions. So, the water used in this process should be demineralized before utilizing it for steam generation. There is a considerable amount of water that is lost in the steam generation process via evaporation. Also, water is lost via transmission from one place to another due to the presence of faulty lines or leaks in the pipeline. Roughly about 1% of the overall water usage is lost because of the above-mentioned problems. Sanitary uses consume about 6% of the overall water usage of the textile industry. Sanitary works include cleaning and washing of the equipment's used for performing various operations in the textile industry. Also, water used for using and cleaning the toilet comes under this fraction of water usage.

15 Future Scope and Limitations

Even though current research is focusing on developing modern technologies to reduce the overall water footprint impact, there is still a very long way to go. The textile industries need to embrace on sustainable development approach such as the use of only organic cotton without any chemicals, a considerable reduction in dyes, and in the production of less complex wastewater which is getting discharged. Governments of respective countries should engage their respective divisions with consumers and business networks to work towards the production of sustainable consumer products. Even while calculating the water footprint for a specific sector, it is very difficult to generalize the data and to give an overview for a conclusive remark, each division of a sector has its potential impact contribution on water footprint by which it makes it difficult for abstraction. Water footprint assessment just focuses on examining freshwater having the primary importance of freshwater scarcity, it does not take into account other environmental factors such as climate change, global warming, soil degradation, and also on other social and economic factors.

16 Conclusion

Water footprint analysis can provide a detailed insight into the water usage of a textile industry which is one of the most polluting industries in the world. Water footprint in a textile majorly stems from the cultivation of cotton and dyeing processes at the industrial level. Water footprint of cotton cultivation varies with the type of agricultural practices adopted with organic farming using significantly less water footprint compared to other methods of farming available. Dyeing process consumes the most water footprint (10–300 L/kg fabric) when compared to other processes involved. Acute estimation of the complete water footprint network of the textile industry can help in minimizing the usage of blue, green, and grey water utilized and simultaneously hindering the water pollution caused by this industry. Although it is tiresome and difficult to account for water footprint for all the processes involved in the textile industry including the labor consumptions, travel consumptions, etc., estimation of water footprint of the processes which consumes the majority of water like wet processing unit and cultivation of raw material can provide a huge boost to the sustainability of the whole textile industry.

References

1. Aivazidou E, Tsolakis N (2019) Water footprint management in the fashion supply chain: A review of emerging trends and research challenges. Elsevier Ltd
2. Aivazidou E, Tsolakis N, Iakovou E, Vlachos D (2016) The emerging role of water footprint in supply chain management: a critical literature synthesis and a hierarchical decision-making

framework. J Clean Prod 137:1018–1037. https://doi.org/10.1016/j.jclepro.2016.07.210

3. Ali M, Sreekrishnan TR (2001) Aquatic toxicity from pulp and paper mill effluents: a review. Adv Environ Res 5:175–196. https://doi.org/10.1016/S1093-0191(00)00055-1

4. Allan JA (1997) "Virtual water": a long term solution for water short Middle Eastern economies? London Sch Orient African Stud Univ London 24–29. https://doi.org/https://dx.doi.org/10.1016/S0921-8009(02)00031-9

5. Antonelli M, Greco F (2015) The water we eat: combining virtual water and water footprints. Water We Eat Comb Virtual Water Water Footprints 1–256. https://doi.org/10.1007/978-3-319-16393-2

6. Babu BR, Parande AK, Raghu S, Kumar TP (2007) An overview wastes produced during cotton textile processing and effluent treatment methods

7. Central Water Commission (2014) Guidelines for improving water use efficiency in irrigation. Domestic & Industrial Sectors, 19

8. Chico D, Aldaya MM, Garrido A (2013) A water footprint assessment of a pair of jeans: the influence of agricultural policies on the sustainability of consumer products. J Clean Prod 57:238–248. https://doi.org/10.1016/j.jclepro.2013.06.001

9. Clift R, Druckman A (2015) Taking stock of industrial ecology. Tak Stock Ind Ecol 1–362. https://doi.org/10.1007/978-3-319-20571-7

10. Dudgeon D, Arthington AH, Gessner MO et al (2006) Freshwater biodiversity: importance, threats, status and conservation challenges. Biol Rev Camb Philos Soc 81:163–182. https://doi.org/10.1017/S1464793105006950

11. Egan M (2011) The water footprint assessment manual. Setting the Global Standard

12. Environment agency (2001) Integrated Pollution Prevention and Control (IPPC) Guidance for the Textile Sector

13. European Commission (2003) Integrated pollution prevention and control reference document on best available techniques for the textile industry. Text Ind 626

14. FAO, ICAC (2015) Measuring sustainability in cotton farming systems

15. Feng L, Dai J, Tian L et al (2017) Review of the technology for high-yielding and efficient cotton cultivation in the northwest inland cotton-growing region of China. F Crop Res 208:18–26. https://doi.org/10.1016/j.fcr.2017.03.008

16. Franke N, Mathews R (2013) Grey Water footprint indicator of water pollution in the production of organic vs

17. Franke N, Matthews R (2013) C&A's water footprint strategy: Cotton clothing supply chain, 58

18. Herendeen RA (2004) Energy analysis and EMERGY analysis—a comparison. Ecol Modell 178:227–237. https://doi.org/10.1016/j.ecolmodel.2003.12.017

19. Hoekstra AY, Chapagain AK (2007) Water footprints of nations: water use by people as a function of their consumption pattern. Water Resour Manag 21:35–48. https://doi.org/10.1007/s11269-006-9039-x

20. Hoekstra AY, Mekonnen MM (2012) The water footprint of humanity. Proc Natl Acad Sci USA 109:3232–3237. https://doi.org/10.1073/pnas.1109936109

21. Hornborg A (2006) Footprints in the cotton fields. Ecol Econ 59:74–81. https://doi.org/10.1016/j.eco

22. Hossain L, Khan MS (2017) Blue and grey water footprint assessment of textile industries of, 437–449

23. Madhav S, Ahamad A, Singh P, Mishra PK (2018) A review of textile industry: Wet processing, environmental impacts, and effluent treatment methods. Environ Qual Manag 27:31–41. https://doi.org/10.1002/tqem.21538

24. Mekonnen MM, Hoekstra AY (2011) The green, blue and grey water footprint of crops and derived crop products. Hydrol Earth Syst Sci 15:1577–1600. https://doi.org/10.5194/hess-15-1577-2011

25. Menezes E, Choudhari M (2011) Pre-treatment of textiles prior to dyeing. Text Dye. https://doi.org/10.5772/19051

26. Muthu Senthilkannan Subramanian (2017) Textile science and clothing technology sustainability in the textile industry
27. Ozturk E, Koseoglu H, Karaboyaci M et al (2016) Minimization of water and chemical use in a cotton/polyester fabric dyeing textile mill. Elsevier Ltd
28. Perry LK (1974) Water resources research institute. Wyo Univ Water Resour Res Inst Water Resour Ser 4823–4839. https://doi.org/10.1002/2014WR016869.Received
29. Pfister S, Boulay AM, Berger M et al (2017) Understanding the LCA and ISO water footprint: a response to Hoekstra (2016) "A critique on the water-scarcity weighted water footprint in LCA." Ecol Indic 72:352–359. https://doi.org/10.1016/j.ecolind.2016.07.051
30. Senthil Kumar P, Grace Pavithra K (2019) Water and textiles. Elsevier Ltd.
31. Senthil Kumar P, Yaashikaa PR (2019) Introduction—water. Elsevier Ltd.
32. Sharma RK (2015) Water and wastewater quantification in a cotton textile industry. Int J Innov Sci Eng Technol 2:288–299
33. Talat F (2018) Synonymous codon usage bias in chloroplast genome of Gossypium thurberi and Gossypium arboreum
34. Wang L, Ding X, Wu X (2013) Blue and grey water footprint of textile industry in China. Water Sci Technol 68:2485–2491. https://doi.org/10.2166/wst.2013.532
35. Wang L, Ding X, Wu X, Yu J (2013) Textiles industrial water footprint: Methodology and study. J Sci Ind Res (India) 72:710–715
36. Wang Z, Huang K, Yang S, Yu Y (2013) An input-output approach to evaluate the water footprint and virtual water trade of Beijing, China. J Clean Prod 42:172–179. https://doi.org/10.1016/j.jclepro.2012.11.007

A Model for the Assessment of the Water Footprint of Gardens that Include Sustainable Urban Drainage Systems (SUDS)

Mª Desirée Alba-Rodríguez, Rocío Ruíz-Pérez, M. Dolores Gómez-López, and Madelyn Marrero

Abstract The limitations presented by traditional urban water cycle systems, which are linearly designed systems, highlight the need to develop new technologies in a new circular strategic approach. In order to quantify the improvements, new methodologies are needed that integrate indicators that assess direct and indirect water consumption, as well as the origin of the water consumed and the incorporation of grey and rainwater. The methodology proposed provides quantitative data in terms of water to calculate the payback period of the new circular systems, comparing the conventional ones with new installations of Sustainable Urban Drainage Systems (SUDS), which are proposed as alternatives to optimize the urban metabolism by improving the water infiltration. The water footprint indicator (WF) is adapted to the construction sector, it allows to quantify the direct and indirect consumption. The first approximation is made to evaluate the impact of the urban water cycle systems. To this end, three possible scenarios are modelled, one of which is a conventional system and another two with SUDS, but different gardens, one of them with autochthonous vegetation and the second one with ornamental vegetation, with greater water requirements. Through this quantification, the amortization period is analyzed in terms of water, considering; the reduction of direct water consumption achieved with the SUDS as compared to the conventional systems; and the consumption of indirect water embedded in the materials necessary for the execution of the systems. The SUDS implementation works require approximately twice as much indirect water as conventional systems, due to the necessary improvements in the terrain for the proper functioning of these eco-efficient systems. This study, together with the technical and economic evaluation, allows us to analyze the viability of the SUDS and contribute with quantitative data in the decision-making phase for the future incorporation of

M. D. Alba-Rodríguez (✉) · R. Ruíz-Pérez · M. Marrero
Higher Technical School of Building Engineering, Architectural Construction Department2, University of Seville, Andalusia, Spain
e-mail: malba2@us.es

M. D. Gómez-López
Higher Technical School of Agricultural Engineering, Agroforestry Engineering Area, Soil and Water Management, Use and Recovery Group, Polytechnic University of Cartagena, Region of Murcia, Spain

© The Author(s), under exclusive license to Springer Nature Singapore Pte Ltd. 2021
S. S. Muthu (ed.), *Water Footprint*, Environmental Footprints and Eco-design of Products and Processes, https://doi.org/10.1007/978-981-33-4377-1_3

this type of eco-efficient systems into the urban networks. The results of the impact of an urban space renovation project applying water-sensitive urban design techniques are shown by evaluating the nature of the materials to be incorporated in the work, the hydrological design of the project, its suitability for the urban environment and its capacity to adapt to future scenarios, evaluating both direct and indirect water. Likewise, the calculation of the WF developed by Hoekstra and Chapagain, generally applied to the agricultural sector, is also adapted to the estimation of the water balance of urban systems with the presence of green areas. The methodology incorporates local biophysical, climatic and temporal data, together with the specific data of the project to calculate the water consumption in the urban area derived from the re-naturalization of urban areas, which has been little explored until now, and to have a measurable indicator to quantify economic and environmental impacts, applicable to the construction sector. In the analysis of the results, it is worth highlighting how the scenarios in which water-sensitive urban design technologies are incorporated presents higher WF values (increased by 1.7 times), referring to the materials and execution of the works than a project in which these design technologies are not applied. The saving of water resources during the use and maintenance phases is 82% per year. The balance means that, at the end of the life cycle, 66% less WF is accumulated and the amortization in terms of water of the infrastructures occurs in year 4.

Keywords Water footprint in urbanization · Sustainable urban systems · Urban systems of sustainable drainage

Glossary

ACCD	Andalusia construction cost database
AP	Auxiliary prices
ARDITEC	Arquitectura: Diseño y Técnica (Architecture: Design and Technique)
ASTM	American Society for Testing and Materials
BP	Basic prices
BREEAM	Building research establishment environmental assessment method
CF	Carbon footprint
CHG	Confederación Hidrográfica Guadalquivir (Guadalquivir Hydrographic Confederation)
CTE	Código Técnico de la Edificación (Building Technical Code)
CUP	Complex unit prices
DC	Direct costs
EF	Ecological footprint
FAO	Food and Agriculture Organization of the United Nations
IC	Indirect costs
IGM	Instituto Geológico Minero de España (Geological Mining Institute)
LCA	Life cycle analysis

LEED	Leadership in energy and environmental design
SUP	Supply prices
UP	Unit prices
WBS	Work-breakdown system
WF	Water footprint
WFblue	Blue water footprint
WFg	Water footprint of the garden
WFgreen	Green water footprint
WFgrey	Grey water footprint
WFN	Water Footprint Network

1 Introduction

The valuation of energy, materials and water and their contribution to the well-being of society is the basis of the current economic system. This makes it necessary to include environmental indicators that allow the impact of any activity to be quantified, thus using the environmental dimension to give a new sustainable approach to the decision-making process in any productive sector. Specifically, according to [19], the construction sector is responsible for 30% of energy consumption and around 40% of natural resources (including water and wood), and produces more than 30% of greenhouse gas emissions.

It is possible to apply various methodologies such as energy analysis [53] and material flow analysis [72], to the construction sector, but the current trend is to use simpler methodologies so that society can easily understand them, and thus enhance their transfer [12].

The Water Footprint Indicator (WF) was developed by Hoekstra [38], for the quantification of water consumed both directly and indirectly by a productive process. Its use is recognized mainly in the sectors of livestock and agriculture, being scarcely employed in the construction sector. Its application to the construction sector will allow the quantification of water consumption associated with construction processes, thus providing an analysis tool to promote the conservation of water resources in the sector.

According to [31], restoring natural water flows, improving water efficiency and preventing pollution, are the strategies on which the current paradigm shift in urban infrastructure and the urban water cycle is based. In order to achieve this new paradigm, the objective is to reach the water balance of the cities [70], which makes it necessary to naturalize the cities. To do this, consolidated areas will require the development of small green masses scattered around the urban environment, which will favour the reduction of local temperatures, not only thanks to the presence of shade, but also because most of the water consumed by plants evaporates, transpires and is returned to the atmosphere, reducing environmental temperatures and incorporating it back into the natural water cycle.

In the particular case of Spain, the most arid country in the European Union, it faces serious challenges in the management of water resources. Spain has one of the largest WFs per inhabitant in the world, amounting to around 6700 L per inhabitant per day. The agricultural sector represents about 80% of total use and the industrial sector 15% of total water use [75].

In order to preserve the environment without damaging the economy of the agricultural sector, a more efficient allocation of water resources is necessary. In this sense, WF analysis can facilitate efficient water allocation and investment, providing a transparent framework for informing and optimizing water policy decisions.

At present, the socio-economic reality is globalized, where people think and act globally without taking into account the particularities of each region or specific area. In the face of this, there is local resistance. From the tensions generated between the global and the local, emerges the concept of Glocalization, a phenomenon recognized and summarized in the literature as "Thinking globally to act locally". It is applied in a variety of contexts, from politics to urban planning, environment, business, culture. It consists of applying global concepts in local actions, [64]. Although efficiency in water resources management is a global problem, the solutions are to be found at the local level. Therefore, it is necessary to have indicators that allow us to know what the starting situation is and to propose possible improvement actions. The WF appears as a response to this need, assessing the water use of organizations, processes and products, and providing quantitative and qualitative information that allows directing efforts towards more sustainable and equitable use of freshwater.

The following sections explain the concepts of WF as an indicator of water resource use; the budgeting system of the Andalusia construction cost database (ACCD), a structure that allows the development of models for environmental impact assessment.

1.1 Water Footprint

The WF is an indicator of water use that is measured in terms of the volume of water consumed and/or polluted per unit of time (m^3/year), it is a geographically and temporally explicit indicator that includes both the direct and indirect use of water for a process, product, consumer and/or producer taking into account all stages of the life cycle. The early works of Lofting and McGauhey [43], calculated volumes of "incorporated" or "embedded" water using input–output analysis [43]. But it was in the early 1990s, that concepts such as water scarcity were developed [23], and J. A. Allan introduced the concept of virtual water, used to calculate the trade balance of a country or territory through its imports and exports [3–5].

The concept of WF was born in 2002, by Professor Arjen Y. Hoekstra, from the University of Twente (Netherlands). Since then, different initiatives have emerged, such as the Water Footprint Network (WFN) in 2008, and ISO 14,046 in 2014, promoting the concept of HH. Its development and standardization came about as a result of the publication of "The Standard Methodology of Calculus" [40] and the "Water Footprint Assessment Manual" (2011) [35].

An WF can be calculated for a particular product, for any well-defined group of consumers (an individual, family, village, city, province, state or nation) or producers (a public organization, private enterprise or an economic sector). Based on the classification made by Mekonnen and Hoekstra in the Water Footprint Assessment Manual (2011) [35]. The WF assesses both direct and indirect water use.

The direct WF, has three components that are differentiated by colour and is related to the consumption of freshwater (blue component), with the risks associated with climate change (green component) and is focused on compliance with discharge regulations, that is, with the quality of ecosystems (grey component). It is, therefore, an effective environmental indicator for understanding how human activities relate to impacts associated with water scarcity and pollution.

The indirect WF is related to eco-design, i. e., the development of products that have a longer life span, that can be reused, repaired, dismantled, as well as fully reused or recycled or their components. It could also differ in its three water components, but this is not often the case. In general, industrial processes are responsible for the release of high volumes of highly polluting chemicals that would require high volumes of water to dissolve if not properly treated. For this reason, the industries that carry out untreated discharges have a high indirect WF due to their grey component, and the industries that treat their discharges properly, complying with the parameters marked by regulations, according to the environmental protection of the receiving waterway, do not calculate their grey component in the indirect WF calculation, reducing the calculation only to the freshwater consumed in the manufacturing process of the material, that is, the blue component [87].

Since its definition and systematization, research in this area has proliferated. The value of this indicator as a tool in decision-making is mainly recognized in the agricultural and livestock production sectors. Preliminary crop estimates have been made at the provincial or national level with explicit spatial data [89], the WF of the agricultural sector and the reservoirs in the Guadalquivir basin has been calculated [69], the relationship between productive sectors (agriculture and livestock) in Andalusia, has been studied in terms of water consumption [86], and also in the industrial sector under a Life Cycle Analysis (LCA) approach [9].

Water is a key resource for the future development of society and advancing knowledge from different perspectives of its management will allow us to improve our understanding of how water governance can be influenced to integrate criteria of environmental sustainability, social equity, economic efficiency and security of supply [39]. The WF is contributing to raising awareness of water issues and its transfer to the urban sector seems appropriate.

To consider WF at the city level, it is necessary to identify and delimit the geographical area of study, identify each of the processes related to urban water supply and sanitation that are carried out in that area, and collect the data on water consumption necessary for each process over a significant period of time. In the urban system, the inputs and outputs to be considered are, on the one hand, the direct inputs which are in the form of rain (green water), groundwater and through the urban water supply system (blue water) and the direct outputs are made through infiltration (blue water), evaporation, plant transpiration (green water), run-off and through the urban sanitation system (grey water). And on the other hand, the indirect inputs referring to the water incorporated in the manufacture of construction materials, and the indirect outputs refer to the water incorporated in the construction waste. Generally, the water inputs and outputs of the system are not constant over time, so depending on the water balance of the system, the water content will change Fig. 1.

In the process of water governance, sufficient technical, administrative and economic capacity has been achieved to ensure basic and adequate supply and sanitation, and even to have an advanced management system in many urban systems in developed countries. From the point of view of urban water management, those solutions that generate a greater reduction in the demand for drinking water, or that reduce the energy consumption linked to the integral management of the urban water cycle, are usually considered to be efficient. However, it is not usual to include in these the LCA linked to these technologies. This work aims to develop, through the calculation of the HH indicator, a model that facilitates decision-making in the field of urban infrastructure related to the urban water cycle.

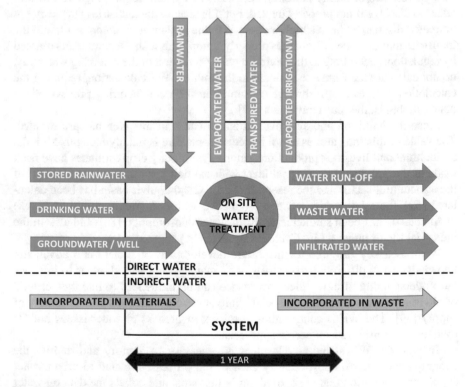

Fig. 1 Water balance of an urban system. Adapt. based on Hoekstra [38]

1.2 Economic Evaluation: Budgeting Model

In the construction sector, the work-breakdown system (WBS) or classification system are the dominant management tools. All these bases are proposed as an ideal tool for the accomplishment of the economic quantification or budgeting and also as an integrating element since its system of decomposition and hierarchization makes it possible to introduce a standardized process. The basic concept in all these systems is to divide a complex problem (project) into increasingly simple parts (units), and these, in turn, are decomposed into the resources needed for its development (basic prices) which are easier to analyze and study, and then all these parts can be added together, without overlap or repetition to define the complex development of the construction project. This cost structure allows a transparent cost estimate to be made at a stage prior to execution, treating the information in a personalized manner according to the objectives to be achieved.

The proposal developed here can be valid for any Construction Cost Base, but, the Systematic Classification used by the Andalusian construction cost database [8] has been used, thus betting on a flexible and adaptable system, with a precise identification of the work units, which allows the uniform elaboration of budgets and which presents a high level of implementation, use and acceptance in the geographical environment of the project. It is used in public housing works by the Andalusian Government and this organization promotes its development and continuous updating. This consists of a model for estimating the costs of projected works by considering execution costs (direct and indirect). An alphanumeric coding allows a precise identification because there is a mandatory biunivocal relationship, that is, each code has only one concept and each concept has only one code.

Direct costs (DC) are all those expenses relating to materials, labour, machinery and installations that are involved in the execution of a specific unit of work and are directly attributable to it by allocating the necessary yields and quantities of each of them. Indirect costs (IC) are all those expenses that cannot be directly allocated to specific units, and which will be passed on by means of an equal percentage of the DC for all units. The ICs include indirect labour dedicated to control and work organization functions, auxiliary means that are not part of a single work unit, the temporary work facilities. Its estimated calculation, in advance, requires a detailed and detailed analysis of the various factors that can influence the determination of the same as the type of work (new plant, recovery), the geometry and immediate environment of the site, the type of building, the programming of its implementation in time, organization and methods of work on site.

Depending on the level of detail sought, the elements of the cost model can be classified by grouping the cost classes into different levels with more or less development in the form of a pyramid, see Fig. 2, in the base are the supply prices (SUP) determined by the manufacturer in accordance with the conditions of supply and receipt of the materials,below are the basic prices (BP) which include together with the SUP the collection and losses of material, distributed mainly according to the three natures mentioned: materials, machinery and labour; in the next level are the

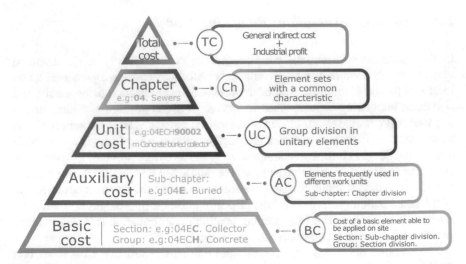

Fig. 2 Cost pyramid structure. Own elaboration

auxiliary prices (AP) formed by the union of BP described above with the quantities appropriate to their type and function, together with the unit prices (UP) formed by the union of BP exclusively or in combination with AP and the complex unit prices (CUP) formed by the union of UP exclusively or in combination with lower level units [62].

All prices are expressed as indicated above, according to a given unit of measurement; but criteria must also be established so that, within the building project in question, the number of units that are subject to that price can be quantified.

1.3 Environmental Evaluation: ArDiTec Model

The various environmental impact reports carried out in recent years highlight the construction sector as one of the main consumers of energy and generators of CO_2 emissions among the various industrial sectors, with estimates of 30–40% of the total environmental impact produced by society (European Parliament-Council of the European Union 2018). This concern has led to the emergence of different types of tools to assess these impacts: through certification and standardization companies, by promoting international standards to encourage the use of environmental labelling of construction products [77, 80–82], the development and application of life cycle analysis (LCA) in this sector [78, 82, 83] and environmental management of buildings from a life cycle perspective [41, 79]. However, the implementation of these standards is not always easy to achieve, due to barriers of all kinds: economic, technical, practical and cultural, which prevent professionals from selecting materials with a lower environmental impact [32, 33].

Most recent studies proposing methodologies for estimating the environmental impact of buildings or the application of ecological indicators to building case studies have focused on aspects such as LCA [10], life-cycle energy consumption analysis [61] or life-cycle carbon footprint [71], or a combination thereof [11, 13]. In recent years, these studies have been incorporated into more powerful computer tools, which is generating a new field of action for LCA, as is the case with BIM platforms [48]. But the LCA methodology and its derivatives are not always easy to use by the non-specialist user, nor is it easy to communicate to such users. Therefore, throughout the development of these indicators, others have been implemented that have a lesser scope, but are easier to use and interpret by the agents involved in construction.

In Spain, there are a variety of these tools that in some way include the calculation of the carbon footprint of buildings. In addition to LEED or BREEAM, whose use has spread in our country, thanks to national bodies such as the Spanish Green Building Council (SpainGBC 2015) and BREEAM Spain (BREEAM 2017). These tools include among the various aspects evaluated to obtain a final score of the CO_2 emissions from the manufacture of building materials and operational energy; however, that final score does not reflect these CO_2 emissions, so it does not report each result separately for better understanding and subsequent analysis of possible improvements.

In the research carried out by Solís-Guzmán (2011), the integration of the Ecological Footprint indicator (EF) in the construction sector is presented, observing the difficulties and benefits that it can generate in relation to other indicators. To this end, first, the indicator must be analyzed to be adapted to the residential sector, by analyzing the construction of buildings, and secondly, a calculation methodology is developed to quantitatively determine the impacts generated by the industry. This methodology applies to the resources used (energy, water, labour, building materials, etc.) and to the waste generated in the construction of residential buildings.

Research has been carried out on the evaluation of the EF indicator of the impact generated by the construction of the 10 most representative typologies of housing constructed in Spain, from 2007 to 2010, identified through official statistical reports [34]. This stage includes the use and maintenance of the building, which is the longest in time, taking into account, on the one hand, direct consumption of water and energy, and on the other hand, those actions of maintenance, conservation and cleaning of the building, necessary to prolong its useful life [49]. A study of the remodelling, rehabilitation and final phase of the building's life cycle, [1, 2] and the HEREVEA project (The Ecological Footprint of Building Recovery: Economic and Environmental Feasibility) which integrates environmental and economic analysis to propose, at the level of Andalusia, aspects related to the application of the HE indicator for decision-making immediately prior to the demolition and end of life of the building [45]. Rivero-Camacho has faced the challenge of analyzing the complete life cycle analysis, a work that has not yet been fully published and whose advances of partial results, referring, for example, to waste [48], arouse interest.

But it is Freire (Antonio Freire [29], who faces the direct environmental costs, similar to the direct economic costs in the project's budget, they cause the direct use of resources in the work through the expenditure of energy of the machinery used in the work (fuel or electricity), [27], through the labour (which entails consumption of food and generation of municipal solid waste, and the consumption of construction materials (energy consumption during its manufacture and implementation) and the indirect environmental costs are all the elements that cannot be attributed to a specific work since they perform tasks that serve several elements simultaneously within the work [25].

The global objectives of sustainable urban development (ODS) [85] are aimed at climate protection and resource conservation. New indicators need to be processed to take into account a differentiating criterion of the committed use of natural resources. Environmental indicators based on LCA are recognized by the scientific community, and as they are easily understood by the society they are given an added meaning. The WF is a geographically and temporally explicit indicator of sustainability, which includes both the direct use of water by a consumer or producer, and the indirect use of water from a process or product.

2   Methodology

Figure 3 shows an outline of the composition of the HH indicator, generally applied to the agricultural sector, based on the definitions developed by Hoekstra and Chapagain [35].

Since this indicator is conditioned by the geographical location and is measured in a time interval, in order to consider WF at the city level, we must identify and delimit the geographical area of study, identify each of the processes related to urban water supply and sanitation that are carried out in that area, and collect the data on water consumption necessary for each process during a significant period of time, which

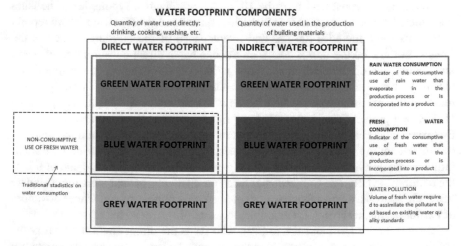

Fig. 3 WF components. Based on Mekonnen and Hoekstra (2011)

will include all stages of water resource demand. In the calculation methodology, this period is a calendar year (WFN) [35].

WF, in turn, for each of the processes, is composed according to its origin in three variables: blue water footprint (WFblue), or freshwater consumed from a surface or underground source extraction; green water footprint (WFgreen), or water from precipitation and that is not converted into runoff and is absorbed by plants; and grey water footprint (WFgrey), or volume of freshwater needed to assimilate the load of pollutants by a receiving body, taking as a reference the environmental quality standards, associating the limits established with good quality for the environment and for people (see Fig. 4, Eq. 1).

The following point presents the model developed for the quantification of the WF in the renewal of urban infrastructures related to the urban water cycle, and not only the processes related to the amount of water necessary for the manufacture of the material resources of construction and its implementation, but also, and in a differentiated manner, the water consumption related to its use directly in the area of study (green areas, irrigation, etc.), taking into account the incorporation of infrastructure designed for water-sensitive areas. This means that the processes taken into account range from the extraction of raw materials, the manufacture of materials, transport, construction and the management of municipal services.

2.1 Model for Quantifying the Direct Water Footprint

2.1.1 Blue Water Footprint (WFblue)

The WFblue is defined by the WFN (see Fig. 4, Eq. 2) and considers the volume of water that is incorporated and evaporates in a process. Most of the time the exact data on the volume of water incorporated or evaporated is not available, so it is posed as a water balance (see Fig. 4, Eq. 3).

The balance of the urban system is conditioned by the flows of inputs (rainfall and irrigation) and outputs (filtration, evaporation and transpiration). Rainfall depends on the local climate; i. e., it is conditioned by the geographical location of the system. Irrigation, in turn, is defined by the water demand of the existing plants within the system. The filtrations are due to the filtration capacity of the underlying soil and the use of systems that allow rainwater to be drained off. Evaporation and transpiration are determined by the presence of plants and the permeability of urban surfaces. Finally, any volume of water that the system cannot assimilate in a given time is transferred to the general sanitation network.

2.1.2 Grey Water Footprint (WFgrey)

In the WFN model, the WFgrey is presented as the volume of freshwater required to assimilate the load of pollutants generated during an activity, based on the water quality standards that are determined by the legislation in force that applies a degree of protection in relation to subsequent uses and the ecological function of the receiving channel where the discharge is located (Fig. 4, Eq. 4).

In the Spanish legislation, the Royal Decree 849/1986, regulates the management and control of the public water domain and the hydrological planning, which determine the seriousness of the infringements in the matter of the public water domain, and establishes the environmental quality standards in the field of the water policy. Their limits are used for the present WFgrey calculations summarized in Table 1. The data on natural pollution have been obtained from the reports of the SAICA Network (Automatic System of Information of Water Quality) for the analysis of the degree of compliance with the parameters analyzed on a continuous basis by the Automatic Analysis Station, with regard to the quality objectives that the Guadalquivir

	nº equation
WF OF A PRODUCTION PROCESS	
WFprocess: Volume of water consumed and/or polluted by a process per unit of time (m³/year)	
$$WF\ process = \sum_{i=1}^{n} (WFblue\ i + WFgrey\ i + WFgreen\ i)$$	1
WF$_{blue\ i}$: Volume of freshwater consumed from surface or groundwater source extraction for the carrying out of a process i per unit of time (m³/year)	
WF$_{grey\ i}$: Volume of freshwater necessary to assimilate the load of pollutants produced in a process i by a receiving body per unit of time (m³/year)	
WF$_{green\ i}$: Volume of water consumed from precipitation that is not converted into runoff for the carrying out o f a process i per unit of time (m³/year)	
Blue Water Footprint	
WF$_{blue}$: Blue water footprint (m³/year)	
$$WF_{blue} = \sum_{i=1}^{n} V_{emb\ i} + V_{evap\ i}$$	2
$$WF_{blue} = VOL_{aff} - VOL_{eff}$$	3
V$_{emb\ i}$: Volume of water incorporated in process i (m³/year)	
V$_{evap\ i}$: Volume of water evaporated in process i (m³/year)	
VOL$_{aff}$: Volume of water entering the system (invoiced) per unit of time (m³/year)	
VOL$_{eff}$: Volume of water leaving the system (consumed) per unit of time (m³/year)	
Grey water footprint	
WF$_{grey}$: Grey water footprint (m³/year)	
$$WF_{grey} = \frac{C_{cont}}{C_{max} - C_{nat}} = \frac{(VOL_{eff} * C_{eff}) - (VOL_{affe} * C_{aff})}{C_{max} - C_{nat}}$$	4
C$_{cont}$: Concentration of pollutant used for the quantification of WF$_{grey}$ (mg/l)	
C$_{max}$: Maximum concentration in the receiving body of the parameter used for the quantification of WF$_{grey}$ (mg/l)	
C$_{nat}$: Natural concentration of the parameter for the quantification of WF$_{grey}$ (mg/l) (without anthropogenic alterations)	
Green water footprint	
WFg: Water Footprint of a garden (m³)	
$$WFg = WFg.\ blue + WFg.\ green + WFg\ grey$$	5
WFg blue: Blue component of the WFg (m³)	
WFg green: Green component of the WFg (m³)	
WFg grey: Grey component of the WFg (m³)	

Fig. 4 WF calculation equations of a production process

Table 1 Limit values of Spanish regulations on dumping for the study area. Sources: R. D. 849/1986 and the river Guadalquivir Hydrographic Confederation (CHG)

Waste water parameter	Units	$C_{máx}$ Table 1 RD	$C_{máx}$ Table 2 RD	$C_{máx}$ Table 3 RD	C_{nat}
Chemical oxygen demand	mg O_2/l	500	200	160	50
Biological oxygen demand	mg O_2/l	300	60	40	100
Total solids	mg/ l	300	150	80	5
Total nitrogen	mg/l	50	50	15	–
Total phosphorus	mg/l	20	20	10	0,1
Total coliforms	ud/100 ml	2,5	2,5	2,5	10
	Source	R.D. 849/1986. Legal limits			CHG

Source Own elaboration based on bibliographical data

Table 2 Most common values of the runoff coefficient [88]

Type of surface	C
Wooded and forest zone	0,10 a 0,20
Zone with dense vegetation granular terrain	0,05 a 0,35
Zone with dense vegetation clayey terrain	0,15 a 0,50
Zone with medium vegetation granular terrain	0,10 a 0,50
Zone with medium vegetation clayey terrain	0,20 a 0,75
No vegetation	0,20 a 0,80
Cultivated zone	0,20 a 0,40
Gravel surface	0,15 a 0,30
Paving	0,50 a 0,70
Bituminous concrete pavement	0,70 a 0,95

Hydrological Plan establishes for the riverbed receiving the discharge a limited water catalogue.

WFgrey can be quantified with different parameters, but when there are several contaminants the WFN standard specifies that the WFgrey of the most significant contaminant should be considered.

2.1.3 Green Water Footprint (WFgreen)

Examples of models, generally applied to the agricultural sector, are known: at the national level [65] based on averages that introduce imprecision into the calculations,at the hydrological basin scale [69] but at the local level, and in particular, for the urban area, there are no representative examples as systematized information on

Table 3 Soil permeability classes [7]

K (cm/s)	Permeability	Soil type		
10^2	Very good drainage	Clean gravel		
10^1				
10^0				
10^{-1}				
10^{-2}	Good drainage	Clean sands	Clays	
10^{-3}		Mixture of gravel and sand	Cracked and altered	
10^{-4}				
10^{-5}	Poor drainage	Fine-grained sand		
10^{-6}		Silts and silty sands		
10^{-7}	Virtually waterproof	Clayey silts (>20% clay)		
10^{-8}		Seamless clays		
10^{-9}				

water consumption in the urban sector does not exist or is incomplete. The methodological basis is that developed by the FAO (Food and Agriculture Organization of the United Nations) [6] incorporating a series of inputs in order to best adapt it to the model of green areas in the urban system and to achieve a better precision in the results.

The sum of the blue, green and grey components allows us to obtain the Water Footprint (WF) produced by the green areas during their growth [35]. By grouping the types of water by colour, it is possible to differentiate the waters according to their quality, access and possibilities of use. The WF of green areas or garden WF is estimated according to Eq. 5 in Fig. 4.

The volume of water consumed by the plants and subsequently transpired, which comes from surface and groundwater sources through irrigation, corresponds to the blue component (WFg blue, m^3/m^2). The volume of water used by the plants, which comes precipitation and stored in the soil is the so-called green component of the garden water footprint (WFg green, m^3/m^2). And finally, the volume of water not used by plants, from precipitation and/or irrigation used, which reaches the urban sewerage system by runoff, is the grey component of garden WF (WFg grey, m^3/m^2). Figure 5 shows the soil water balance, as well as the flow in and out of the soil moisture in a rain garden, based on the quantification previously developed for crops (Arjen Y. [37].

Rainfall

An urban basin hydrological study, unlike the river basin study, requires taking into account a number of particularities. The main one is the dimensions, much smaller than those corresponding to the rivers. Among the various calculation procedures that exist for the flow of rainwater is the rational method that is applied by the Spanish

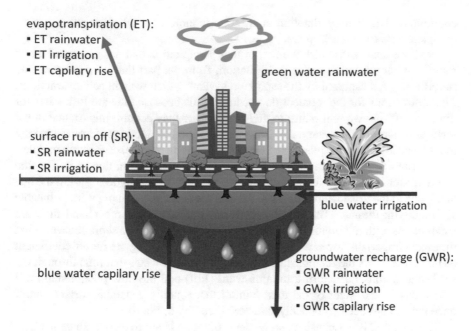

evapotranspiration (ET):
- ET rainwater
- ET irrigation
- ET capilary rise

green water rainwater

surface run off (SR):
- SR rainwater
- SR irrigation

blue water irrigation

blue water capilary rise

groundwater recharge (GWR):
- GWR rainwater
- GWR irrigation
- GWR capilary rise

Fig. 5 Soil water balance: flow in and out of the soil moisture in a rain garden. Adaptation based on [37]

standard instruction of roads, 5. 2-IC, (Ministerio [55]. More simply, it is the curve number method that was empirically established by the Soil Conservation Service (SCS. USA) [56], which is the most generalized and has been used to simulate the behaviour of rain and runoff in a basin for many years, its implementation in a software for the formulation of projects for small river basins (TR-55), modified for HEC-HMS the HMS model [22], are the most accepted.

Design rainfall or total rainfall (PTo) is obtained from globalized information extracted from CLIMWAT 2.0, a climate database published by FAO's Water Development and Climate Change and Bioenergy Management Unit. Its choice is marked by the fact that it also offers observed agroclimatic data from more than 5,000 stations worldwide. In addition to monthly rainfall (mm/month), the database provides long-term average values of 6 more parameters: average daily maximum temperature (°C), average daily minimum temperature (°C), average relative humidity (%), average wind speed (km/day), average hours of sunshine per day, average solar radiation (MJ/m^2/day) and effective monthly rainfall (mm/month). It also incorporates a tool that provides the reference evapotranspiration (ETo) calculated with the Penman–Monteith method (mm/day).

For the time series, CLIMWAT [15], to provide reliability in its data includes historical data from direct observations available from the 1971–2000 period, including all recent series that have at least 15 years of data, even if they are not continuous. The average duration of a series is more than 50 years and the most complete ones exceed 200 years.

Next, we must distinguish the part of that precipitation that has generated direct runoff, which we call effective precipitation, from the part that does not generate runoff and is not captured by the sanitation system, which we will call abstractions. The abstractions for the moment do not interest us because they are lost water for direct runoff, but we will return to them later since they acquire importance in the integral balance of the system since one part evaporates, another infiltrates soaking the soil and is used by the plants, and the rest remains on the surface forming puddles.

The runoff produced in the specific case of an urban basin, due to its specific characteristics, is very different from that originating in rural hydrology. On the one hand, due to the urbanization process, the surface is altered, making it impermeable and modifying the natural network of runoff and drainage. On the other hand, these are small basins with minimum infiltration surfaces, which produce short concentration times and important flows. It can be simplified by calculating the runoff coefficient (C) that represents the relationship between the flow of water that runs through the surface as a consequence of a rainfall event (PEf) and the flow precipitated on it (PTo), that is, the water of the total rainfall that actually generates surface runoff once the soil has been completely saturated (see Fig. 6, Eq. 6).

This factor C is not constant, so its determination is approximate, since it varies according to the magnitude of the rain and the topographic, soil and land use conditions. A sloping area does not respond to the same amount of rain as a plain, nor does a garden area in front of a road. The runoff coefficient can take values between zero and one ($0 \leq C \leq 1$). They are determined from the annual values of precipitation and flow rate and are tabulated in the literature on the subject of surface hydrology. The following table presents the most common C values used for the calculation of urban basins (see Table 2).

In any case, the determination of an average run-off coefficient for a uniform area can be considered. And when an area is made up of different types of land, different surfaces are calculated as the average run-off coefficient by making a weighted average of the different run-off coefficients of each of the sub-areas into which the area under consideration can be divided. In this way, we arrive at the expression of the weighted runoff coefficient (see Fig. 6, Eq. 7) for an area formed by different sub-areas with different runoff coefficients.

Based on the fact that the unit of litres/m^2 is equivalent to the precipitation in mm, the volumes of water are deducted. In other words, one litre spread over a surface area of one square metre produces a sheet of water one millimetre thick. On the one hand, if we apply the surfaces considered to PTo, we obtain the total volumes of precipitation (VTo), and on the other hand, if we apply them to the effective precipitation (PEf), we obtain the volumes of runoff (VEf). The difference is the volume of abstractions or losses (Vab) (see Fig. 6, Eqs. 8–10).

The infiltrations

The permeability of soil is its capacity to allow the flow of a fluid through it. This mechanical property of soils, which is measured by the coefficient of permeability

(K), is a constant, so it is assumed that infiltration is constant throughout the storm, and is expressed in terms of speed (cm/s). Darcy (1856) defined this parameter for sands and showed that it was equally valid for other soils. The methods for both field and laboratory determination are standardized by the American Society for Testing and Materials (ASTM). From Darcy's law, the expression that relates the flow of water through a sample to its permeability taking into account the pressure differential (Fig. 6, Eq. 11) is derived.

When measuring infiltration values, Bárcena and Hurtado [7], mention that the range of permeability coefficient K values is very wide and extends from 102 cm/s for very coarse gravel, to a negligible value in the case of clays, as presented in Table 3.

A portion of the precipitation that falls on the earth infiltrates into the soil and becomes part of the groundwater. The volume of water filtered through a surface is related to the effective rainfall, the permeability of the soil through the infiltration flow and the permeable or infiltration section or surface (Fig. 6, Eq. 12). Once on the ground, some of this water moves close to the earth's surface and quickly emerges to be discharged into the beds of the water currents, but due to gravity, a large part of it continues to move towards deeper areas.

The movement of water below the surface depends on the permeability and porosity of the subsurface rock. If the rock allows water to move relatively freely within it, water can move significant distances in a short period of time. But water can also move to deeper aquifers, from where it will take years to become part of the environment again.

Water demand

For the quantification of the real evapotranspiration (water demand) of crops and green areas, the procedure proposed by FAO is used, which introduces the concepts: reference evapotranspiration and the crop coefficient (Fig. 6, Eq. 13). Reference evapotranspiration is calculated with a class A cube for each area, as this is a valid low-cost method for estimating high diffusion and evapotranspiration for a large number of production plots [21].

In the formulation of the water demand calculation, a correction coefficient is introduced (crop coefficient (Kc)) that contemplates the biophysical variations of the plant, that is, how the height of the plant and soil cover vary in the growth cycle, since this influences the evapotranspiration. Which means that this coefficient varies over time. Depending on the phase in which the plant is, germination, growth, flowering and ripening (from sowing to harvesting), the amount of water consumed by the crop varies. This variation is also reflected in the productivity of the crop, depending on the sizes and maximum quantities.

Evapotranspiration of a reference crop

In order to calculate the water demand of a crop, it is necessary to know the effect of the climate at a certain point to obtain an ideal crop. This data is called reference evapotranspiration (ETo), expressed in millimetres of height of evapotranspired water per day (mm/day). As its name suggests, it is used as a reference indicator and is

	nº equation
WATER FOOTPRINT OF A GARDEN	
RAINFALL	

$$PEf = PTo * C$$

6

$$C = \frac{\sum_{i=1}^{n} S_i * C_i}{\sum_{i=1}^{n} S_i}$$

7

PEf: annual effective rainfall (mm/year)

PTo: average annual rainfall (mm/year)

C: weighted runoff factor of an area (non-dimensional)

Si: area of sub-area i (m^2)

Ci: average coefficient of sub-area i (non-dimensional)

Precipitation volume

$$VTo = \frac{PTo * S}{1000}$$

8

$$VEf = \frac{PEf * S}{1000}$$

9

$$VAb = VTo - VEf$$

10

VTo: Total precipitation volume (m^3)

VEf: Effective precipitation or runoff volume (m^3)

VAb: Abstraction or loss volume (m^3)

PTo: average annual rainfall (mm/year)

PEf: annual effective rainfall (mm/year)

S: Area of the basin or surface considered (m^2)

Infiltration

$$Qi = K * I * A$$

11

$$VIn = \frac{PEf \ x \ Qi \ x \ S}{1000}$$

12

Qi: quantity of water drained through a sample per unit of time (cm^3/h)

K: coefficient of permeability or hydraulic conductivity (cm/h)

I: available piezometric gradient (m/m)

A: cross section through which water infiltrates the sample (cm^3)

VIn: Volume of infiltrated water (m^3)

Water demand

$$ETc = Kc \ x \ ETo$$

13

$$ETg = Cg \ x \ ETo$$

14

$$Cg = Fs \ x \ Fd \ x \ Fm$$

15

ETc: Crop evapotranspiration

Kc: Coefficient specific to each crop

ETo: Evapotranspiration of a reference crop

ETg: Garden evapotranspiration

Cg: Specific coefficient for each garden

Fs: Species factor

Fd: Density factor

Fm: Microclimate factor

Fig. 6 Equations to calculate the WF of a garden: precipitation, infiltration and water demand

estimated through meteorological data of the place where the crop is implanted. The method used to obtain these reference values (ETo) has been the Penman–Monteith Method, considered the most scientifically valued [76], as it allows for unrestricted use in any type of climate. According to the CROPWAT program developed by the FAO [14], which calculates the irrigation requirements of crops based on climate, crop and soil data, it establishes the need to resort to irrigation if the amounts of water provided by rain do not cover the needs to ensure the survival of the crop.

The ETo is obtained from the relative humidity, the maximum and minimum air temperature, the sunshine and the wind speed. These climatic data needed for the calculation are measured from weather stations, equipped with humidity, radiation, wind speed and temperature measurements. The data used in the calculation have been extracted from the statistics developed by agencies and/or institutes, using the meteorological data collected by the stations closest to the study area. Specifically, data are extracted from the FAO CLIMWAT database [14].

Adapting the crop coefficient to the garden coefficient

Due to the very different conditions that exist between the field crop areas and the urban gardens and green areas covered in this paper, it is not possible to apply directly the crop coefficients extracted from the CROPWAT database based on data from FAO publications. There are three main differences between the farmland and the urban garden areas; firstly, related to the location and the environment, since in these garden areas a microclimate effect is generated due to their relative situation with the surrounding buildings, which influences aspects such as the predominant winds and the generation of shade, other aspects also influence, such as road traffic and paved areas; secondly, due to the differences in species located within the same garden space, with different levels of species such as shrubs, trees or herbaceous being found simultaneously without the predominance of one over the other and all with different irrigation needs; and finally, related to the lack of uniformity in the density of vegetation, with greater soil evaporation in areas where the density of vegetation is lower.

On the basis of all this, the garden coefficient (Cg) method is used, which is based on the method described above, introducing the necessary modifications to specify the losses that occur in the garden areas (garden evapotranspiration (ETg)) (Fig. 6, Eq. 14).

In the case of the gardens, the concept of "maximum productivity" applied to the areas of cultivation is not applied, this concept is changed to "adequate aesthetics", thus achieving the optimum results with the minimum contribution of water, quantities that are much lower than those needed in the crops. For this reason, in gardening, a correction coefficient called the garden coefficient (Cg) will be applied, which allows the calculation of the amount of water necessary for the gardens to obtain an adequate aesthetic, that is to say, an adequate growth and healthy appearance. To take into account these conditions, the species factor (Fs), the density factor (Fd) and the microclimate factor (Fm) appear (Fig. 6, Eq. 15). The garden coefficient is not a constant, since it reflects the variations in the water needs of the plants throughout their development, so this value will vary through the different stages of plant growth, from germination, through growth, flowering and pruning, etc. It is considered an average garden coefficient obtained for a period of one year Fig. 7.

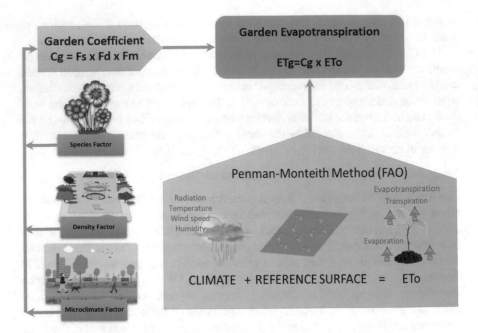

Fig. 7 Flow of the determination of garden evapotranspiration. Own elaboration

Table 4 Species factor [17]

Species factor (Fs)	Minimum	Maximum
Very low	0.01	0.10
Low	0.11	0.30
Moderate	0.31	0.60
High	0.61	0.90

Species Factor

The species factor includes the calculation of the water requirements taking into account the characteristics of each botanical species. In Table 4, the values for the different species can be observed depending on the needs. Two ranges are distinguished, the Fs values included within the very low range, which will be applied in the cases of an ideal situation for the use of xero-gardening, and the Fs values included within the high range, which will be used for species that require greater contribution of water to maintain aesthetics.

To obtain the data, the publication made by Cotello and his team (1999), was used, in which more than 1800 species are listed, III Manual WUCOLS (Water Use Classification of Landscape Species) [17]. When choosing the value of Fs, it must be taken into account that this database refers to climatic zones of California, and this means that it may have a different value of Fs if it is located in our climatic zone of

Table 5 Density factor [16]

Density factor (Fd)	Minimum	Maximum
Low	0.20	0.70
Medium	0.71	1.00
High	1.01	1.30

study or another, because the same species depending on the temperature can vary its water needs [16].

Normally, there will be a combination of different species with similar water needs in the same place, so it will act in the same way as in the previous case since the PIFs of the species will be the same or very similar. In the case of an area with several species and different water needs, which is not the normal case of design, a value slightly lower than the highest Fsi should be used, to avoid damaging the species with lower needs and that do not suppose a significant decrease in the species with the greatest demand for water. To calculate the Fs of the set, the Fsi of each species are weighted using the number of individuals as specific weight.

Density Factor

The density factor takes into account the grouping of plants for the calculation of water needs, so that the evaporation rate is higher the higher the plant mass density. The values shown in Table 5, range from the highest values, when plants are present in high density grouped in various levels of cover, to the lowest values, when species are in isolation. Three levels of overlap are considered: upper or arboreal, middle or shrubby and lower or herbaceous, coinciding with the classification defined by the FAO in the Land Cover Classification System (LCCS) [20]. It is considered high if there is an overlap of the three levels and low when there is no overlap of levels.

Microclimate Factor

As indicated above, the density factor depends on urban aspects such as the location of adjacent buildings and the shade they cast on the landscaped areas, the protection that the buildings can offer from the prevailing winds, the presence of heat sources such as road traffic and pavements. The range of this coefficient is between 0. 40 at its lowest value (protected areas and areas with shade from buildings and other plants) and 1. 40 at its highest value (areas exposed to heat sources such as high traffic density, excessive sunshine, etc.) (see Table 6).

Table 6 Microclimate factor
[16]

Microclimate factor (Fm)	Minimum	Maximum
Low	0.50	0.99
Medium	1.00	1.00
High	1.11	1.40

Effective Rainfall

For the calculation of the actual rainfall for plants, the volume of abstractions (Vab) is considered, the portion of rainwater lost to the direct runoff, whose destination was varied, one part evaporates, another amount of infiltrates soaking the soil and is used by the plants, and the rest remains on the surface forming ponds. Rigorously calculating the part of the VAb used by the plants can become very complex, which is why simplified methods are usually used. In this case, it has been chosen to use the simple curve number (NC) method of the Soil Conservation Service/Natural Resources Conservation Service of the USDA of the SCS (P), [57].

To calculate the irrigation needs of a garden, the FAO method is used, in which the natural water balance of the study area is determined by considering the existing water gains and losses in the garden. This takes into account gains from rising water, runoff and losses of water due to deep percolation, but in the case of the gardening calculation, the established simplifications mean that these values are considered negligible. This simplification is supported by the very definition of a well-designed irrigation system, i. e., with almost no deep percolation and no runoff. Therefore, the losses produced by evapotranspiration must be compensated by the sum of irrigation water and the rainfall, in order to achieve the water balance of the system (Fig. 8, Eq. 16).

The precipitation water used by the plant (RWA- Rain Water Available), is the one that through the effective precipitation of rain (PEf) begins to satisfy in the first place the demand of water requested by the plant as a consequence of the evapotranspiration (ETg) for a determined period. The excess rainfall for the plant (RWS-Rain Water Surplus) is that which is produced when the difference between evapotranspiration and rainfall is positive, i. e., there is excess rainwater, so this amount of water is unusable, remaining in the soil and even influencing the withering of the plants [42]. This surplus is converted into percolated water (grey water), providing an opportunity to use this surplus as a resource to satisfy other water needs. On the other hand, there are periods in which there is a deficit or water needs in the garden (GWR-Garden Water Requirement), this occurs when the difference between evapotranspiration and rainfall is negative, so these requirements should be covered with irrigation. Calculations of these requirements can be done monthly, every fifteen days or every twelve days.

Calculated the natural water balance of the system, the next step is to study how to enhance the irrigation of each hydro-zone (area with similar water needs), this study should consider the needs of plants to evolve and be maintained optimally. In the calculation of these gross water requirements (Nb), it must be considered that there

	nº equation
WATER FOOTPRINT OF A GARDEN	
EFFECTIVE RAINFALL	

$$ETg = PEf + GWR \qquad 16$$

Corollary:

If ETg ≤PEf	then	RWA=ETg	and	RWS=PEf-ETg	and	GWR=0
If ETg >PEf	then	RWA=PEf	and	RWS=0	and	GWR=PEf-ETg

ETg: Garden evapotranspiration
PEf: Effective rainfall in the area (mm)
GWR: Garden irrigation water requirement (mm)
RWS: Rain water surplus(mm)

Localised irrigation

$$If\ Ea \le (1 - Fl)\ then: \quad Nb = \ GWR/Ea \ x \ 100 \qquad 17$$

$$If\ Ea > (1 - Fl)\ then: \quad Nb = \frac{GWR}{1 - Fl} \ x \ 100 \qquad 18$$

Sprinkler irrigation

$$If\ Fl \le 10\%\ then: \quad Nb = \frac{GWR}{Ea} \ x \ 100 \qquad 19$$

$$If\ Fl > 10\%\ then: \quad Nb = \frac{09x\ GWR}{1 - Fl} \ x \ 100 \qquad 20$$

Ea: Irrigation application efficiency factor (in points to one)
Nb: Gross water requirement (mm)
Fl: Washing fraction (in points to one)

Irrigation Water Surplus

$$IWS = \ GWS - Nb \qquad 21$$

IWS: Irrigation water surplus (mm)

Components of the garden's WF

$$WFg\ green = \ S \ x \sum_{i=1}^{n} RWAi \ x \ Si \qquad 22$$

$$WFg\ blue = \ S \ x \sum_{i=1}^{n} GWRi \ x \ Si \qquad 23$$

$$WFg\ grey = \sum_{i=1}^{n} (RWSi + IWSi) \ xSi \qquad 24$$

RWA: Rain water available (mm)
S: Area considered (m²)
Si: área of sub-area i (m²)

Fig. 8 Equations to calculate the WF of a garden: effective rainfall

are a series of external factors that influence the measurement and quantification of these requirements and that can vary the need for extra water inputs, these factors are: the presence of salts in the soil and the fact that the irrigation systems are not fully efficient.

In order to know the efficiency of the irrigation system, the method used and its effectiveness is of great importance, since the percentage of water used by the roots of the plants depends to a great extent on this. In order to quantify this efficiency, both filtrations, run-off and evaporation of irrigation water are considered, depending on the irrigation system used, all through the application efficiency coefficient (Ea).

Table 7 Efficiency factors in the application of irrigation water [16]

Efficiency app. type of irrigation (Ea)	Minimum	Maximum
Irrigation located underground	0.95	0.95
Irrigation located on the surface	0.90	0.90
Diffusers and micro-sprinklers	0.80	0.80
Sprinklers	0.70	0.80
Surface	0.50	0.65

The coefficients expressed in Table 7, are intended to serve as a guide for knowing the efficiency of the irrigation method used. Considering the total water supplied and the percentage of water used by the roots of the plants, different levels of efficiency of the systems are achieved, obtaining losses of only 5% in underground irrigation systems compared to losses of up to 50% in surface irrigation systems.

In the calculation of the irrigation needs, it is also necessary to consider another variable called washing fraction (Fl), this variable allows to know the extra irrigation water needs that must be provided in the function of the soil salinity and the irrigation water salinity. The values of the least salt-tolerant species should always be chosen for application. Once the washout values and the application efficiency of the system are obtained, the raw values are obtained from Eqs. 17 to 20 in Fig. 8.

As indicated above and as can be seen from Eq. 11 in Fig. 8, the irrigation water not used by the plant (IWS-Irrigation Water Surplus) results in runoff.

Equations 22 to 24 in Fig. 8, break down all the elements required for the calculation of the different components (blue, green and grey) of the garden water footprint (WFg).

The direct WF of the system (Fig. 4, Eq. 5), is applied once the three components have been quantified. To do this, the three objectives to be achieved in order to achieve a more sustainable system must be taken into account: to stop the waste of "blue water", to make better use of "green water" and to tend to zero in "grey water", therefore, the blue and grey components whose tendency will be reductionist for the optimization of the system, and therefore, have the opposite sign to the green component that will tend to grow to improve the balance of the system.

The methodology described above is a valid tool for calculating the water footprint of urban gardens. As can be seen in Fig. 9, the model includes all the parameters considered necessary for the calculation, with a precision obtained thanks to the literature consulted, as well as the existing gardening bases, which allows the methodology to be used in different scenarios to estimate and evaluate water requirements, and thus serve as a tool for the adoption of measures to improve urban gardening systems.

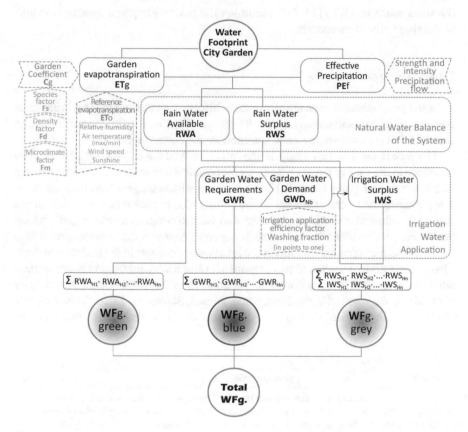

Fig. 9 Flowchart; determination of the green, blue and grey water footprint of the city garden (WFg). Own elaboration

2.2 Model for Quantifying the Indirect Water Footprint

In this case, the theoretical framework is divided into two parts: methodology of economic analysis of the project and adaptation of this methodology to face environmental analysis. Like the Ecological Footprint (EF) methodology of Solís-Guzmán (J. [73], it starts from an economic analysis and has been adapted to measure the complete life cycle of the building: urbanization (Madelyn [47], use and maintenance [51], 2016), and rehabilitation or demolition [2].

The building's LCA is completed and advanced, including the study of new indicators such as incorporated energy (EE) [26], carbon footprint (CF) [28, 73, 74] and the water footprint (WF) [24, 66], becoming the most widespread indicators thanks to the simplicity of its concept.

2.2.1 Economic Analysis

Data and process automation are advances in information technology that bring great advantages in predictive analysis. In the sector, WBS or construction information classification systems are the most common management tools.

The model used in this study is the WBS of Andalusia [44], which has been described in the state of the art, based on a hierarchical and tree structure with defined levels, where each group is divided into subgroups with homogeneous characteristics. The classification system materializes the codification of each concept, which means that each code corresponds to a concept and each concept to a unique code, which allows the precise identification of each concept. Among other advantages, it also facilitates its management and solves the location of concepts in the budget structure. This organization of work and its components offer a very robust and stable system when it comes to dividing a complex system such as the work budget into simpler elements, i. e., materials, machinery and labour. Shown in Table 8, an example of PU from the gardening chapter. All of the above characteristics facilitate the

Table 8 Example of gardening price

15JAA00120	u	Slow-growing shade tree	Total €	81.01

Slow-growing, decorative special evergreen tree 2.50 m high served bare rooted, including today's opening 1 × 1 m. soil extraction, planting and filling of topsoil, supply of biological fertilizers, 2 m high chestnut wood stake, conservation and watering. Measure the quantity executed

Code	Quantity	U	Description	Price	Cost
ATC02100	0.500	h	Masonry crew, formed by first class experts, assistant specialist and worker specialist	38.75	19.38
UJ00165	1,200.00	kg	biological fertilizer	0,02	24.00
UJ00300	1.000	u	Slow-growing, evergreen tree 2,50 m	16.16	16.16
UJ01800	1.000	m^3	Topsoil	8.37	8.37
UJ01900	2.000	u	chestnut wood stake 2 m	5.91	11.82
MK00100	0.050	h	dump truck	25.60	1.28

Source Own elaboration

incorporation of environmental cost based on the same hypotheses and contours defined in the calculation of economic cost.

2.2.2 Environmental Analysis: Indirect Water Footprint

The environmental calculation model developed by the ARDITEC Group to evaluate all the stages of the building's life cycle allows the evaluation of different environmental indicators [24]. In our case, the starting document will be the project's budget and the environmental impact of the consumption of the resources consumed is evaluated by means of a methodology based on the analysis of the LCA, using international databases of LCA of construction products [52]. In order to facilitate the introduction of materials with lower environmental impact into the construction sector, this methodology for calculating the footprint of materials takes into account the energy required during the life cycle of the cradle to door. From the general analysis of the different UP that makes up the construction budget, the direct costs (DC) are extracted, which are broken down into machinery, labour and materials Fig. 10.

Machinery

The energy consumption of machinery is considered to be the main environmental impact, linked to the power of its engine and to working hours (Madelyn [47]. For

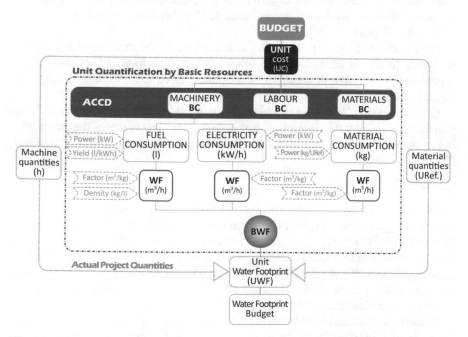

Fig. 10 Indirect water footprint calculation model based on ARDITEC methodology. Own elaboration

the calculation of WF a distinction must be made between electrical machinery and machinery powered by an engine fuelled by fossil fuels, either diesel or petrol. The calculation of WF is made based on engine power, in the same way as it is done in other studies [24].

The average fossil fuel consumption of machinery is linked to engine power. For its calculation, the "Manual de maquinaria" drawn up by SEOPAN (2008), is used, which collects technical data on different models and typologies of machinery on the market, arriving at the elaboration of the following calculation hypothesis by choosing the most unfavourable consumptions:

Diesel: 0.15 to 0.20 L consumed in 1 h per kW installed.

Petrol: 0.30 to 0.40 L consumed in 1 h per kW installed.

The machinery classified in the BCCA is analyzed, where the previous coefficient is applied to the power of each engine to obtain the litres of fuel consumed per working hour, differentiating whether the machine consumes diesel or petrol, and obtaining the average consumption per hour of machinery that consumes either diesel or petrol (Fig. 11, Eq. 25).

To the average consumption of each construction machinery, the conversion factor of the volume of water needed to have one kilo of fuel is applied, extracted from the open database Ecoinvent [30], implemented in Symapro and developed by the Swiss Centre for Life Cycle Inventories, due to its transparency in the development of processes (reports, flowcharts, methodology, etc.), consistency, references and standing out for the fact that it merges data from several databases of the construction industry (Martínez-Rocamora, Solís-Guzmán, [47]. For this case, it has a value of 1.26 m^3/kg (Fig. 11, Eq. 26).

For electrical machinery, the power consumption of the machinery is analyzed according to the engine power. The water consumption factor is applied for the production of one kWh of energy by the Spanish electrical system [63], (Fig. 11, Eq. 27) which in this case has a value of 0.0118 m^3/kWh.

Once the unitary WF of the machinery consumption of the work is obtained, the project measurement (Qi) is applied, that is to say, the time of use of the machinery in the work of work and the total HH of the machinery of the project is obtained (Fig. 11, Eq. 28).

Construction materials

The first step to be taken in order to obtain the WF indicator for each element consists of converting the original unit of measurement of each PB (m^3, m^2, metres, tonnes, thousands, etc.) to Kg (Fig. 11, Eq. 29), so that we can apply the density established in the support documents used, the Catalogue of Construction Solutions of the Technical Building Code (CSC) and the Basic Document on Structural Safety of the Technical Building Code. Actions in the Building DB-SE AE (RD 314/2006), to obtain the weight of each element.

Once the weight has been obtained, the LCA databases will be used, which define the units of impact contained in each kg of material. Among the different LCA databases, the Ecoinvent database was chosen [30], implemented in Symapro and developed by the Swiss Centre for Life Cycle Inventories, due to its transparency in

	nº equation
INDIRECT WATER FOOTPRINT	
Machinery	
$$V = \text{Pot} \times \text{Rend}$$	25
$$WF_{comb\ i} = V_i \times D_{comb\ i} \times F_{WF\ comb}$$	26
$$WF_{elec\ i} = V_i \times F_{WF\ elect}$$	27
$$WFB_{Maq} = \sum_{i=1}^{n} (WF_{comb\ i} \times Q_i) + (WF_{elec\ i} \times Q_i)$$	28
V: average fuel consumption of the machinery (l/h)	
Pot: power of the machinery's engine (kW)	
Rend: fuel consumed by the machinery's engine depending on whether it is diesel or petrol (l/kWh).	
WF$_{comb\ i}$: basic water footprint of the fuel consumption (fossil) of machinery i (m³/h)	
Vi: average fuel or electricity consumption of machinery i (l/h) or (kWh)	
D$_{comb\ i}$: average fuel density of machinery i	
F$_{WFcomb}$: fuel water consumption factor (m³/kg)	
WF$_{elec\ i}$: basic footprint of the electrical consumption of the machinery i (m³/h)	
F$_{WFelec}$: water consumption factor of the energy mix (m³/kWh)	
WFB$_{Maq}$: basic water footprint of machinery (m³)	
Q$_i$: measurement of machine use i (h)	
Construction materials	
$$V = U \times F_{URef}$$	29
$$WF_{m\ i} = V_i \times F_{WF\ mi}$$	30
$$WFB_{Mat} = \sum_{i=1}^{n} (WF_{m\ i} \times Q_i)$$	31
V: unit of material consumed expressed in weight (kg)	
U: unit of material expressed in its unit of reference (URef)	
F$_{URef}$= factor for conversion from reference unit to weight of the unit of material (kg/URef)	
WF$_{m\ i}$: basic water footprint of material i (m³/URef)	
V$_i$: unit of material i (kg)	
F$_{WFm\ i}$: water consumption factor of material i (m³/kg)	
WFB$_{Mat}$: basic water footprint of material (m³)	
WF$_{m\ i}$: basic water footprint of material i (m³/URef)	
Q$_i$: measurement of material consumption i (m³/URef).	

Fig. 11 Calculation equations for the indirect WF

the development of processes (reports, flowcharts, methodology, etc.), consistency, references and standing out for the fact that it merges data from several databases of the construction industry (Martínez-Rocamora, Solís-Guzmán, [47]. From this database, a series of "environmental families" have been obtained which will be responsible for assigning to each BP their corresponding impact basic according to their similarity.

Of the life cycle inventory for each of the materials, the reference used for calculating indirect WF is that disseminated by the Water Footprint Network (WFN), whose Hoekstra [38] concept and calculation methodology "The Standard Calculation Methodology" and "The Water Footprint Assessment Manual" (A. Y. Hoekstra, Chapagain, Aldaya, & Mekonnen, 2011). To then calculate the direct and indirect consumption of any productive process expressed in volume of water consumed (m³/Kg) (Fig. 11, Eq. 30).

Once the basic WF of the material consumption of the work is obtained, the project measurement (Qi) is applied, that is, the quantities consumed in the work and the total WF of the materials is obtained (Fig. 11, Eq. 31).

3 Scenarios

Three scenarios are modelled, one of them provided with the so-called conventional (S1) and another two with SUDS, in which two typologies of garden areas are incorporated, one of them with native vegetation (S2) and the other with ornamental vegetation (S3), with higher water requirements. The area of action is the same for all scenarios and affects a developed area of 2,781 m^2.

The territorial enclave is a street in Seville, therefore, the local climate data correspond to the weather station No. 1381 [15], (see Fig. 12, a warm and temperate climate, with a high level of climatic welfare during much of the year, average annual temperature of 19.2 °C. Mild winters, the coldest month is January, and frosts are rare, temperatures do not tend to fall below 2.0 °C. Warm summers, the hottest month is July, and the highest average temperatures are 36.0 °C. The mild and pleasant weather, with plenty of sunshine reaching an average of 354 h of sunshine a year. The rainfall regime and its average distribution is 539 mm/year. The driest month is July, with an average rainfall of 2 mm and an average relative humidity of 44% and the wettest month is December, with an average rainfall of 97 mm and an average relative humidity of 74%,the presence of severe phenomena is rare. 50.5 is

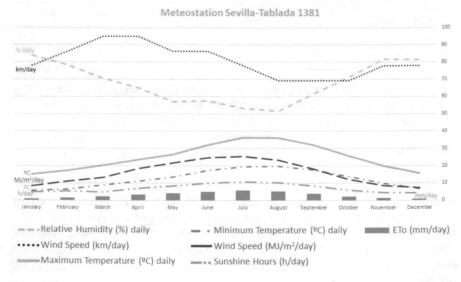

Fig. 12 Monthly average values of daily data obtained from Sevilla-Tablada Meteorological Station [14]

the average number of days of rain per year that exceeds 1 mm and 9. 1 is the average number of storm days per year.

The terrain of the urban area, hydrologically the aquifer of Seville, locally has its water table around 4 m deep. The water-bearing materials are calcarenites from the Upper Miocene and different alluvial materials from the Quaternary terraces of the Guadalquivir. This profile allows the groundwater to be fed into the ground by direct infiltration of rainwater, since the area is classified as medium permeability by the Spanish Geological Mining Institute (IGME), corresponding to a Quaternary detrital basin, so the use of this data (3.3×10^{-4} cm/s).

S1 consists of a renovation of the supply and sanitation networks where the street is restored to its previous state, transforming the urban space into an impermeable platform that directs all rainfall and runoff to the general sanitation system together with the water discharged from homes and businesses. As for the materials incorporated in the work, it involves the installation of a supply network of 75 m of 250 mm diameter ductile iron pipe together with its corresponding valves, hydrants, connections to the network in service and replacement of the current lead home connections with others made of polyethylene. A sewerage network of 140 m of vitrified stoneware pipe in diameters 500 and 600 with their corresponding manholes, overhang, manholes and scuppers. And the replacement of the affected roads to their initial state, being the roadway a bituminous pavement and the pavements of ceramic paving stones and prefabricated concrete kerbs.

S2 and S3, or garden typologies, show a SUDS that manages 2,781 m^2, consisting of a sanitation system disconnected from the existing one, with two infiltration areas of 152 and 153 m^2 on both sides of the road Fig. 13 and 540 m^2 of rain gardens. The precipitated water collected from the roads and pavements is led to both bioretention areas by its upper part, it is filtered by the filler material to the bottom where it stores a maximum of 86 m^3. The nature of the terrain also allows for its infiltration into the ground. The rest of the water runs off through a drainage pipe that drains into a 45 m^3 cistern. In the event of torrential rains, the spillways are connected to the existing urban sanitation system.

In addition to the materials already described in S1 for the general supply and sanitation installations, the materials derived from the execution of the SUDS are improvements to the land by means of gravel and aggregate fillings, PVC drainage pipe. The reinforced concrete cistern, with its corresponding electromechanical

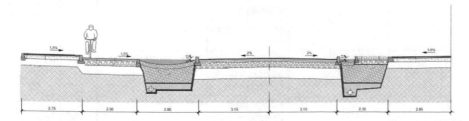

Fig. 13 Street profile for S2 and S3. Own elaboration

equipment and the necessary valves to supply the polyethylene irrigation network installation formed by a general network of diameter 63, a mesh of drip pipes and a general network of support diffusers.

S2 with the implantation of native species consists of a low-density shrub (lantana camara) plantation established in a semi-shaded area by existing consolidated trees (citrus aurantium, jacaranda mimosaefolia and celtis australis). S3 consists of a mixed plantation of shrubby Mirato rose and trees (fraxinus angustifolia, prunus cerasifera). Both are alternated with a mixed Kikuyo and English ray-grass pavement-plant at 50%, considering at the level of calculation of the garden coefficient Table 9, the differentiation in three hydro-zones, which is defined as a delimited garden area containing a series of plants with similar water needs, and therefore, an overall water behaviour.

From the natural balance derived from the rainfall available in the area and the water requirements of the group of plants that make up each scenario, the annual irrigation requirements for each hydro-zone are deducted. The hydro-zone of S2 has

Table 9 Application of the garden coefficient calculation method

Hydro-zone		Species or type of vegetation	Fsi	Fs	Fd	Fm	Cg
E2	H1 Autochthonous deciduous trees	Citrus Aurantium var. amara (1)	0.50	0.27	1.00	1.40	0.378
		Jacaranda Mimosaefolia (1)	0.50				
		Celtis Australis (1)	0.42				
		Lantana camara (3)	0.26				
E3	H2 Ornamental rosebush	Jacaranda Mimosaefolia (11)	0.50	0.50	1.10	1.30	0.715
		Fraxinus angustifolia (3)	0.50				
		Prunus Cerasifera var. pissardii (2)	0.40				
		Shrubby Mirato rose (1,5)	0.50				
E2 and E3	H3 Grasses	English Ray-grass 50% Lolium hybrid	0.60	0.40	1.00	0.80	0.320
		Kikuyo 50% Pennisetum clandestinum	0.20				

a less water-demanding native profile than that of S3 with an ornamental profile Figs. 14 and 15.

Integrating the natural system of hydro-zones in the urban system composed of a SUDS, storage in a cistern and use for irrigation of the rainfall in the area; the direct WF is quantified in each of the scenarios for a normal year. The appreciable differences between not having adopted a project for the renovation of urban facilities by means of a solution that includes water-sensitive urban design elements (S1) and others in which it does (S2 and S3), distinguishing its components according to their colour: blue, green and grey, see Table 10.

For each scenario, the indirect WF has been evaluated following the environmental budgeting methodology, and the water impact of the materials incorporated in the work. To do this, the basic resources of the project are identified, and a database of

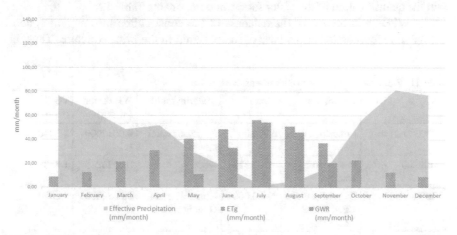

Fig. 14 Natural balance of S2, autochthonous hydro-zone

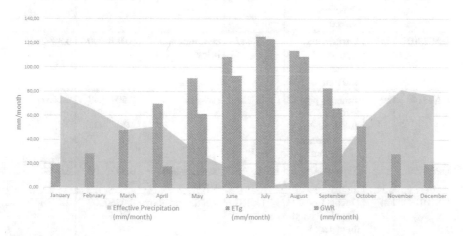

Fig. 15 Natural balance of S3, ornamental hydro-zone

Table 10 Annual direct water footprints with distribution by type of water

Scenarios	S1	S2	S3
WF d blue (m^3)	82.18	648.49	771.58
WF d green (m^3)	0.00	531.07	583.87
WF d grey (m^3)	1.561.39	635.84	596.72

unitary elements is created, whether they are machinery or material, and conversion factors are applied to them according to their nature, thus obtaining the basic WF of each element. An example is taken from each family of materials in Table 11.

Based on the quantification of the initial scenario (S1), the input and output volumes of each of the scenarios described have been evaluated, obtaining the input and output flows of each system in terms of total volumes for an average year together with the quantification of the water footprint components Table 12.

The WFBs will be part of the database of the construction information system and will be incorporated at the base of the pyramid in the price structure. Thus, in

Table 11 Extract from the basic water footprint database

Code	Ud	Description	Factor	Material family	WF Factor (m^3/kg)	WFB (m^3/URef)
AC00300	m^3	Limestone filler	2,770.000	aggregates and stones	0.0012000	3.3240
CM00200	m^3	Pinewood board	509.850	wood	0.0026245	1.3381
GC80040	t	Bagged cement CEM II/B-P 32,5 R	1,000.000	Concrete and cement	0.0030000	3.0000
MC00400	h	Sandblaster	1.500	machinery	1.2600000	1.8900
MR00500	h	Vibratory Concrete Screed	0.480	machinery	1.2600000	0.6048
UA02303	m	Vitrified clay pipe, rated diameter 150 mm	25.947	ceramics and bricks	0.0150000	0.3892
UA03140	m	drainage pipe PVC of double-wall, rated diameter of 200 mm	12.227	plastics	0.5043647	6.1667
UP01450	t	Asphalt mix type S-12	1,000.000	bitumen and asphalt	0.0030000	3.0000
US10109	m	Ductile iron pipe: K-9 rated diameter of 150 mm	18.860	metals and alloy	0.0230000	0.4338

Table 12 Inputs, outputs and direct annual WF

Scenarios	S1	S2	S3
Inputs			
Total precipitation volume (Vto) m^3	1,643.57	1,643.57	1,643.57
Well water m^3	0.00	177.75	308.59
outputs			
Garden evapotranspiration (ETg) m^3	0.00	531.07	583.87
Volume of infiltrated water (VIn) m^3	0.00	457.72	457.72
Drainage volume m^3	1,561.39	338.43	338.43
WF components			
Blue (m^3)	0.00	160.01	289.94
Green (m^3)	0.00	531.07	583.87
Grey (m^3)	1,561.39	635.84	596.72

order to elaborate the auxiliary WF, the unit WF or the complex unit WF in the same way that the AP, PU, PUC were elaborated, the quantities of project material for each one of the cases were applied, how in the price table of an economic budget, one starts from the simplest units to obtain the most complex units. An example of unitary WF is given in Table 13: m^3 of concrete: HA-30/P/20/IIa + Qb in floors and foundations. Finally, by applying the project's measurements to the unitary WFs, the project's WF results are obtained.

Table 13 Extract from the table of unitary water footprints

03HAL00725	m^3	CONCRETE HA-30/P/20/IIa + Qb IN FLOORS AND FOUNDATIONS

Reinforcement concrete HA-25/P/20/IIa, plastic consistency, and maximum aggregate size 20 mm, in slabs and foundations, supplied and installed, also p. p. of bottom cleaning, vibrating and curing, according to EHE and CTE (Building Technical Code). Measuring the theoretical executed volume

Code	Ud	Quantity	Description	WFB (m^3/URef)	WF (m^3/URef)
TP00100	h	0.240	Specialist Worker	-	-
CH80020	m^3	1.000	Concrete HA-30/P/20/IIa,	5.1439	5.1439
MR00500	h	0.080	Vibratory Concrete Screed	0.6048	0.0484
				TOTAL	**5.1923**

4   Results

The results of the impacts of the materials incorporated into the work are presented in Table 14, expressed in the unit of measurement of the environmental indicator per unit of the urban area affected.

Housing developments, like domestic buildings, offices, infrastructure or any other product, have a useful life, that is, their materials are exhausted or their facilities lose their functionality. When faced with the doubt of delimiting this period, town planners usually have as a reference annex III of Royal Decree 1492/2011, of 24 October, which approves the valuation regulations of the Land Law [54], which states that roads, open-air car parks and similar have a specified useful life of 40 years. However, these terms can vary depending on the materials used, it is not the same new work as a work of 40 years ago,it also depends on its functionality since it can precipitate its obsolescence of use; and also much influence is had by the maintenance and investments that are made, it is not the same performance of "patching" that a comprehensive performance.

With the methodology developed, it is possible to obtain the total WF of each scenario Table 15, expressed in volumes of water per urban surface and from whose results the amortization period in terms of water is analyzed and compared, considering,the reduction of the water impact achieved with the SUDS, during the use and maintenance phases, is 82% per year, compared to the direct water consumption of the conventional systems. The balance means that the amortization in terms of water of the infrastructures occurs in year 4, since the end of the life cycle, 66% less WF is accumulated.

The most interesting aspect of the indicator is its simple concept which is easy to calculate and to understand by the public. Another powerful aspect is that it allows the aggregation of factors from different sources into a single indicator and a total impact assessment. On the downside, the aggregation is based on assumptions on data about the potential pollution of fertilizers or street dirt, the precipitation frequency, plants water necessities that are not obtain empirically for the precise garden location, adding uncertainties. However, the WF can also be studied per category (blue, green and grey), which helps the most impacting aspects to be identified in each.

Table 14 Indirect water footprints per surface urbanized

Scenarios	S1	S2	S3
Indirect WF (m^3/m^2)	2.68	4.61	4.58

Table 15 Total Water Footprint per surface area at the end of the life cycle

Scenarios	S1	S2	S3
Indirect WF (m^3/m^2)	2.68	4.61	4.58
Direct annual WF (m^3/m^2)	0.56	0.10	0.11
Direct WF (m^3/m^2) of life cycle	22.40	4.00	4.40
Total WF	**25.08**	**8.61**	**8.98**

The indicator has been used since its inception to determine the impact of the agricultural sector but can be also useful to assess the urbanization. This methodology, adapted to the unique characteristics of the construction sector, has been chosen for its comprehensibility, transparency, and adaptability.

5 Conclusions

The indirect WF of the scenarios that apply a water-sensitive urban design (S2 and S3) is 1.7 times greater than the scenario in which only the water supply and sanitation facilities are renewed and the street is restored to its original state (S1). Local rainwater management in scenarios S2 and S3 improves the direct WF of S1 by 82% per year. This saving at the end of the life cycle accumulates 66% less footprint, it is in year 4, when these projects accumulate a total WF of 5.00 m^3/m^2 and are equal in water consumption, and therefore, the amortization in terms of water of the infrastructures is produced.

The incorporation of SUDS has made it possible to reduce the volume of peak rainfall relieved from the sewage network by 78% in its area, leading to energy savings by reducing the volume of pumping and wastewater treatment that the municipal system should treat. Including infrastructures that take advantage of rainwater for irrigation, in addition to offering better management of urban runoff, reducing the overloading of sewage networks and the dragging of pollutants, allows alternative water resources to be available by being able to recirculate around 362 m^2 of rainwater, for the conservation of green spaces, and therefore, less dependence on general supply systems. The use of high-efficiency irrigation systems is confirmed as a water-saving measure.

The incorporation of green areas and irrigation infrastructures, the incorporation of plants with a lower water demand or native plants (S2), as opposed to those with a higher water demand or ornamental plants (S3), although the general balance is similar, the difference in the green WF is higher in ornamental plants, and therefore, represent an improvement in the biophysical environment; ornamental plants require more irrigation and the blue component increases, even despite the use that the system makes thanks to the reuse of rainwater; on the other hand, native plants, as they do not make use of rainwater, the grey component is higher, hence in both cases the balance is similar. An optimal scenario would be an urban garden where areas are distributed with plants with different water requirements that consume rainwater in an optimal way, minimize the demand for irrigation water and tend to zero the unused water.

6 Future Lines of Research

To implement the calculation of indirect WF as an environmental criterion for evaluating offers within public procurement processes. In the construction sector, the public sector has a prominent role in the path towards a more sustainable production system, since with its power of choice it can prioritize the purchase of certain brands or products with better environmental performance.

To measure the sensitivity to climate change and therefore evaluate impacts and vulnerability of systems related to the urban water cycle. This is possible because the green component of the WF indicator is linked to the geographical area of study and to a given time period, which makes it possible to determine the most appropriate species for the amount of water resource available in that area and at any given time, and by incorporating future local climate data obtained by projecting changes in the climate for different gas emission scenarios.

Include the evaluation of the quality of the outside environment, analyzing other parameters such as temperature, humidity, sunshine, sound levels and other atmospheric pollutants to estimate the improvements in environmental comfort produced by the implementation of water-sensitive solutions in urban space.

To transfer the concept of WF to that of EMergy Synthesis defined by Odum [58–60] which will mean the transition from Environmental Economics to Ecological Economics allowing the evaluation of economic and natural systems, as well as the interactions between both with a common methodology.

Acknowledgements The work has been partially supported by Cátedra del Agua, EMASESA.

References

1. Alba-Rodríguez MD, Marrero M, Solís-Guzmán J (2013) Economic and environmental viability of building recovery in seville (Spain) phase 1 : database in arcgis. Cathedra Chair of Housing Environmental. Faculty of Architecture Cracow University of Technology, 11/2013, pp 297–302
2. Alba-Rodríguez MD, Martínez-Rocamora A, González-Vallejo P, Ferreira-Sánchez A, Marrero M (2017) Building rehabilitation versus demolition and new construction: economic and environmental assessment. Environ Impact Assess Rev 66:115–126. https://doi.org/10.1016/j.eiar. 2017.06.002
3. Allan JA (1993) Fortunately there are substitutes for water otherwise our hydro-political futures would be impossible. Priorit Water Res Alloc Manag 13(4):26
4. Allan JA (1994) Overall perspectives on countries and regions. In: Rogers P, Lydon P (eds) Water in the Arab world: perspectives and prognoses. Harvard University Press, Cambridge, MA, pp 65–100
5. Allan JA (1998) Virtual water: a strategic resource. Ground Water 36(4):545–547
6. Allen RG, Pereira LS, Raes D, Smith M (1998) Crop evapotranspiration-guidelines for computing crop water requirements-FAO Irrigation and drainage paper 56. Fao, Rome 300(9):D05109
7. Bárcena YA, Hurtado JEA (2011) ENSAYOS DE PERMEABILIDAD EN MATERIALES DE BAJA DE PERMEABILIDAD COMPACTADOS
8. BCCA (2017) Consejería de Fomento y Vivienda / Vivienda y Rehabilitación / Base de Costes de la Construcción de Andalucía (BCCA). Retrieved from https://www.juntadeandalucia.es/fomentoyvivienda/portal-web/web/areas/vivienda/texto/7a0899c8-0038-11e4-8cc4-27ee69 a25823
9. Berger M, Finkbeiner M (2010) Water footprinting: how to address water use in life cycle assessment? Sustainability 2(4):919–944

10. Buyle M, Braet J, Audenaert A (2013) Life cycle assessment in the construction sector: a review. Renew Sustain Energy Rev 26:379–388. https://doi.org/10.1016/j.rser.2013.05.001
11. Cabeza LF, Rincón L, Vilariño V, Pérez G, Castell A (2014) Life cycle assessment (LCA) and life cycle energy analysis (LCEA) of buildings and the building sector: a review. Renew Sustain Energy Rev 29:394–416. https://doi.org/10.1016/j.rser.2013.08.037
12. Cagiao J, Gómez B, Doménech JL, Gutiérrez Mainar S, Gutiérrez Lanza H (2011) Calculation of the corporate carbon footprint of the cement industry by the application of MC3 methodology. Ecol Ind 11(6):1526–1540. https://doi.org/10.1016/j.ecolind.2011.02.013
13. Chau CK, Leung TM, Ng WY (2015) A review on life cycle assessment, life cycle energy assessment and life cycle carbon emissions assessment on buildings. Appl Energy 143(1):395–413. https://doi.org/10.1016/j.apenergy.2015.01.023
14. CLIMWAT (2018a) CLIMWAT 2.0. FAO databases an software. Retrieved from https://www.fao.org/land-water/databases-and-software/climwat-for-cropwat/en/
15. CLIMWAT (2018b) CLIMWAT 2.0. FAO Databases an Software. https://www.fao.org/land-water/databases-and-software/climwat-for-cropwat/en/, consultado en mayo de 2018
16. Contreras F, González A, López J, Calvo A (2006) Estimación de necesidades hídricas para especies de jardín en la región de Murcia: adaptación de WUCOLS y utilización del sistema de información agraria de Murcia. III Jornadas Ibéricas de Horticultura Ornamental, Almería, Spain
17. Costello LR, Jones KS (2000) Water use classification of landscape species (WUCOLS III). A guide to estimating irrigation water needs of landscape plantings in California. Department of Water Resources, Sacramento, California
18. CROPWAT (2018) CROPWAT 8.0. FAO databases an software. https://www.fao.org/land-water/databases-and-software/cropwat/en/, consultado en mayo de 2018
19. Cuchí A, Wadel G, Rivas P (2010) Sector edificacion: La imprescindible reconversion del sector frente al reto de la sostenibilidad
20. Di Gregorio A (2005) Land cover classification system. Classification concepts and user manual. Software version 2
21. Doorenbos J, Pruitt WO (1977) Crop water requirements. FAO irrigation and drainage paper 24. Land and Water Development Division, FAO, Rome, p 144
22. Engineers USACO (2000) Hydrologic modeling system HEC–HMS, technical reference manual. Hydrologic Engineering Center, Davis, CA
23. Falkenmark M (1989) Water scarcity and food production. Food and Natural Resources, pp 164–191
24. Freire-Guerrero A, Alba-Rodríguez MD, Marrero M (2019) A budget for the ecological footprint of buildings is possible: a case study using the dwelling construction cost database of Andalusia. Sustainable Cities and Society 51:101737. https://doi.org/10.1016/J.SCS.2019.101737
25. Freire-Guerrero A, Marrero-Meléndez M (2015) Ecological footprint in indirect costs of construction. In: Proceedings of the II International congress on sustainable construction and eco-efficient solutions : Seville 25–27 may 2015 (pp. 969–980). Retrieved from https://www.mendeley.com/research-papers/ecological-footprint-indirect-costs-construction/?utm_source=desktop&utm_medium=1.19.1&utm_campaign=open_catalog&userDocumentId=%7Bbfc310cc-56d5-4362-82b9-8720e901d018%7D
26. Freire-Guerrero A, Marrero M (2015) Evaluation of the embodied energy of a construction project using the budget. Habitat Sustentable 5(1):54–63
27. Freire A, Marrero M (2014) Analysis of the ecological footprint produced by machinery in construction. In World Sustainable Building 14. Barcelona
28. Freire Guerrero A, Marrero M, Muñoz Martín J (2016) Incorporación de huella de carbono y huella ecológica en las bases de costes de construcción. Estudio de caso de un proyecto de urbanización en Écija, España. Hábitat Sustentable, ISSN-e 0719–0700, vol. 6, no. 1, 2016 (Ejemplar Dedicado a: Junio), pp 6–17, 6(1), 6–17
29. Freire Guerrero A. (2017). Presupuesto ambiental. Evaluación de la huella ecológica del proyecto a través de la clasificación de la Base de Costes de la Construcción de Andalucía. Tesis Doctoral. Universidad de Sevilla

30. Frischknecht R, Jungbluth N, Althaus H-J, Doka G, Dones R, Heck T, … Rebitzer G (2005) The ecoinvent database: overview and methodological framework (7 pp). Int J Life Cycle Assess 10(1):3–9

31. Gaffron P, Huismans G, Skala F (2008) Proyecto ecocity. Manual Para El Diseño de Ecociudades En Europa. Libro I La Ecociudad: Un Lugar Mejor Para Vivir. Bilbao: Centro Documentación Estudios Para La Paz

32. Giesekam J, Barrett JR, Taylor P (2016) Construction sector views on low carbon building materials. Build Res Inf 44(4):423–444. https://doi.org/10.1080/09613218.2016.1086872

33. Giesekam J, Barrett J, Taylor P, Owen A (2014) The greenhouse gas emissions and mitigation options for materials used in UK construction. Energy Build 78:202–214. https://doi.org/10.1016/J.ENBUILD.2014.04.035

34. González-Vallejo P, Solís-Guzmán J, Llácer-Pantión R, Marrero-Meléndez M, Llácer R, Marrero M (2015) La construcción de edificios residenciales en España en el período 2007–2010 y su impacto según el indicador Huella Ecológica. Informes De La Construcción 67(539):e111. https://doi.org/10.3989/ic.14.017

35. Hoekstra AY, Chapagain AK, Aldaya MM, Mekonnen MM (2011) The Water Footprint Assessment Manual

36. Hoekstra AY, Chapagain AK, Aldaya MM, Mekonnen MM (2011) The Water Footprint Assessment Manual

37. Hoekstra AY (2019) Green-blue water accounting in a soil water balance. Adv Water Resour 129:112–117. https://doi.org/10.1016/J.ADVWATRES.2019.05.012

38. Hoekstra AY (2003) Virtual water trade between nations: a global mechanism affecting regional water systems. IGBP Global Change News Letter 54:2–4

39. Hoekstra AY (2017) Water footprint assessment: evolvement of a new research field. Water Resour Manage 31(10):3061–3081

40. Hoekstra AY, Chapagain AK, Aldaya MM, Mekonnen MM (2009) Water Footprint Manual State of the Art 2009, (November)

41. ISO 15686–5 (2017) Buildings and constructed assets—Service life planning—Part 5: Life-cycle costing

42. Lanza Espino G. de la, Martínez C, Martínez A, Pulido H (1999) Diccionario de hidrología y ciencias afines

43. Lofting EM, McGauhey PH (1968) Economic evaluation of water-part IV: an input-output and linear programming analysis of California water requirements

44. Marrero M, Ramirez-De-Arellano A (2010) The building cost system in Andalusia: application to construction and demolition waste management, 28 Construction Management and Economics §, Routledge. https://doi.org/https://doi.org/10.1080/01446191003735500

45. Marrero M, Martin C, Muntean R, González-Vallejo P, Rodríguez-Alba MD (2018) Tools to quantify environmental impact and their application to teaching: projects City-zen and HEREVEA. IOP Conf Series: Mater Sci Eng 399:012038. https://doi.org/10.1088/1757-899x/399/1/012038

46. Marrero M, Rivero-Camacho C, Alba-Rodríguez MD (2020) What are we discarding during the life cycle of a building? Case studies of social housing in Andalusia, Spain. Waste Manage 102:391–403

47. Marrero M, Puerto M, Rivero-Camacho C, Freire-Guerrero A, Solís-Guzmán J (2016) Assessing the economic impact and ecological footprint of construction and demolition waste during the urbanization of rural land. Resour Conserv Recycl 117:160–174. https://doi.org/10.1016/j.resconrec.2016.10.020

48. Marrero M, Wojtasiewicz M, Martínez-Rocamora A, Solís-Guzmán J, Alba-Rodríguez MD (2020) BIM-LCA integration for the environmental impact assessment of the urbanization process. Sustainability 12(10):4196

49. Martinez-Rocamora, Alejandro, Solís-Guzmán, Jaime, Marrero-Meléndez M (2015) A structure for the quantity surveillance of costs and environmental impact of cleaning and maintenance in buildings. In: The sustainable renovation of buildings and neighbourhoods. Bentham Science Publishers, pp 103–118. Retrieved from https://ebooks.benthamscience.com/book/9781681080642/chapter/131503/

50. Martínez-Rocamora A, Solís-Guzmán J, Marrero M (2016) Toward the ecological footprint of the use and maintenance phase of buildings: utility consumption and cleaning tasks. Ecol Ind 69:66–77. https://doi.org/10.1016/j.ecolind.2016.04.007

51. Martínez-Rocamora A, Solís-Guzmán J, Marrero M (2017) Ecological footprint of the use and maintenance phase of buildings: maintenance tasks and final results. Energy Build 155:339–351. https://doi.org/10.1016/j.enbuild.2017.09.038

52. Martínez-Rocamora A, Solís-Guzmán J, Marrero M, Marrero-Meléndez M (2016) LCA databases focused on construction materials: a review. Renew Sustain Energy Rev 58:565–573. https://doi.org/10.1016/j.rser.2015.12.243

53. Meillaud F, Gay J-B, Brown MT (2005) Evaluation of a building using the emergy method. Sol Energy 79(2):204–212

54. Miguel CMG (2011) Real Decreto 1492/2011, de 24 de octubre, por el que se aprueba el Reglamento de valoraciones de la Ley de Suelo (BOE núm. 270, de 9 de noviembre). Actualidad Jurídica Ambiental 8:17–18

55. Ministerio de Fomento (2016) Norma 5.2-IC" Drenaje Superficial" (Orden FOM/298/2016 de 15 de febrero). MOPU, Madrid (Spain)

56. Mockus V (1964) National engineering handbook. Section

57. NRCS U (2004). Estimation of direct runoff from storm rainfall. National Engineering HandbookÀPart, 630

58. Odum HT (1996) Environmental accounting: emergy and environmental decision making. Wiley

59. Odum HT (1998) Emergy evaluation. In: Odum HT (1998, May) Emergy evaluation. International workshop on advances in energy studies: energy flows in ecology and economy. Porto Venere, Italy

60. Odum HT (2002) Emergy accounting. In Unveiling Wealth. Springer, pp 135–146

61. Ramesh T, Prakash R, Shukla KK (2010) Life cycle energy analysis of buildings: an overview. Energy Build 42(10):1592–1600. https://doi.org/10.1016/j.enbuild.2010.05.007

62. Ramírez-de-Arellano-Agudo A (2014) Presupuestación de obras. Editorial Universidad de Sevilla (2014). Sevilla

63. REE (2014) El sistema eléctrico español/The Spanish electric system

64. Robertson R (1995) The search for fundamentals in global perspective. In: The search for fundamentals. Springer, pp 239–262

65. Rodriguez Casado R, Garrido Colmenero A, Llamas Madurga MR, Varela Ortega C (2008) La huella hidrológica de la agricultura española. Papeles De Agua Virtual 2:1–38

66. Ruiz-Pérez MR, Alba Rodríguez MD, Marrero M (2017) Systems of water supply and sanitation for domestic use. In I. I. C. on C, B. Research (ed.) Water footprint and carbon footprint evaluation: first results. Santa Cruz de Tenerife, España

67. Ruiz-Pérez MR, Alba-Rodríguez MD, Castaño-Rosa R, Solís-Guzmán J, Marrero M (2019) Herevea tool for economic and environmental impact evaluation for sustainable planning policy in housing renovation. Sustainability 11(10):2852. https://doi.org/https://doi.org/10.3390/SU11102852

68. Ruiz-Pérez MR, Alba-Rodríguez MD, Marrero M (2019) The water footprint of city naturalisation. Evaluation of the water balance of city gardens. In: The 22nd biennial conference of The International Society for Ecological Modelling (ISEM). Salzburg, Austria.

69. Salmoral G, Dumont A, Aldaya MM, Rodríguez-Casado R, Garrido A, Llamas MR (2012) Análisis de la huella hídrica extendida de la cuenca del Guadalquivir. Fundación Marcelino Botín

70. Sarté SB (2010) Sustainable infrastructure: the guide to green engineering and design. John Wiley & Sons

71. Schwartz Y, Raslan R, Mumovic D (2018) The life cycle carbon footprint of refurbished and new buildings—a systematic review of case studies. Renew Sustain Energy Rev 81:231–241. https://doi.org/10.1016/j.rser.2017.07.061

72. Sinivuori P, Saari A (2006) MIPS analysis of natural resource consumption in two university buildings. Build Environ 41(5):657–668

73. Solís-Guzmán J, Martínez-Rocamora A, Marrero M (2014) Methodology for determining the carbon footprint of the construction of residential buildings. Springer Singapore, pp 49–83. https://doi.org/https://doi.org/10.1007/978-981-4560-41-2_3
74. Solís-Guzmán J, Rivero-Camacho C, Alba-Rodríguez D, Martínez-Rocamora A (2018) Carbon footprint estimation tool for residential buildings for non-specialized users: OERCO2 project. Sustainability 10(5):1–15. https://doi.org/10.3390/su10051359
75. Sotelo Navalpotro JA, Olcina Cantos J, García Quiroga F, Sotelo Pérez M (2012) Huella hídrica de España y su diversidad territorial
76. Souza JM de, Pereira LR, Rafael A, da M., Silva LD da, Reis EF. dos, Bonomo R (2014) Comparison of methods for estimating reference evapotranspiration in two locations of Espirito Santo. Revista Brasileira de Agricultura Irrigada 8(2):114–126. https://doi.org/https://doi.org/10.7127/rbai.v8n200225
77. UNE-EN 15804 (2012) Sustainability of construction works—environmental product declarations—core rules for the product category of construction products
78. UNE-EN 15978 (2012) Sustainability of construction works. Assessment of Environmental Performance of Buildings, Calculation Method
79. UNE-EN ISO 14001 (2015) Environmental management systems—requirements with guidance for use
80. UNE-EN ISO 14020 (2002) Environmental labels and declarations—general principles
81. UNE-EN ISO 14021 (2017) Environmental labels and declarations—self-declared environmental claims (Type II environmental labelling)
82. UNE-EN ISO 14025 (2006) Environmental labels and declarations—type III environmental declarations—principles and procedures
83. UNE-EN ISO 14040 (2006) Environmental management—life cycle assessment—principles and framework
84. UNE-EN ISO 14044 (2006) Environmental management—life cycle assessment—requirements and guidelines
85. United Nations (2012) UN general assembly resolution on the future we want (adopted on 27 July 2012), (A/RES/66/288)
86. Velázquez E (2006) An input–output model of water consumption: analysing intersectoral water relationships in Andalusia. Ecol Econ 56(2):226–240
87. Wang L, Ding X, Wu X (2014) Careful considerations when reporting and evaluating the grey water footprint of products. Ecol Ind 41:131–132
88. Woodward SM, Posey CJ (1955) Hydraulics of steady flow in open channels
89. Zeng Z, Liu J, Koeneman PH, Zarate E, Hoekstra AY (2012) Assessing water footprint at river basin level: a case study for the Heihe river basin in northwest China. Hydrol Earth Syst Sci 16(8):2771–2781

From Field to Bottle: Water Footprint Estimation in the Winery Industry

Melody Blythe Johnson and Mehrab Mehrvar

Abstract The Water Footprint (WF) concept allows the quantification of the impact of the production of a wide range of consumer products on local water resources. Each stage in the manufacturing and distribution process adds complexity and variability in the WF estimation. It is, therefore, critical to clearly define the scope of the footprint assessment and any underlying assumptions made. The wine-making process is a complex mix of agricultural, chemical, sanitizing, bottling and distribution processes, each exerting unique demands on local water resources. While wine production is recognized as a water-intensive process, clear guidelines for determining the blue, green and grey WF are essential to allowing the benchmarking and comparison of WF values from location to location. Factors affecting the WF can include local climate conditions, vineyard irrigation practices, pesticide and fertilizer use, soil quality, proximity to surface and groundwater resources, manufacturing process efficiency, water reuse practices, and wastewater handling and treatment methods and performance. Case studies quantifying the WF of wine manufacturing are available in the published literature; however, the lack of a standardized industry-wide WF assessment approach makes benchmarking and comparing the performance of winery operations difficult. A framework for the assessment of the WF specific to wine-making is necessary to accurately quantify water resource impacts, benchmark performance and prioritize the implementation of mitigation measures to reduce environmental impacts.

Keywords Benchmarking · Framework · Wine production · Wineries · Water footprint

M. B. Johnson · M. Mehrvar (✉)
Department of Chemical Engineering, Ryerson University, 350 Victoria Street, Toronto, ON M5B2K3, Canada
e-mail: mmehrvar@ryerson.ca

© The Author(s), under exclusive license to Springer
Nature Singapore Pte Ltd. 2021
S. S. Muthu (ed.), *Water Footprint*, Environmental Footprints and Eco-design
of Products and Processes, https://doi.org/10.1007/978-981-33-4377-1_4

103

1 Introduction

Wine production represents a significant worldwide industry, with total volumetric production in 2019 of 260 million hectoliters of wine [29]. In terms of economic impact, the wine-making industry contributed an estimated €14 billion to the economy of Italy alone in 2018 [45], a country long recognized for wine production. Even in smaller wine-producing regions, the local economic impacts can be significant. For example, in 2012 Canada was home to a modest 476 wineries employing over 3,700 employees and revenues of over $1 billion, with an annual average growth rate of over 5% (Agriculture and Agri-Food Canada, nd). Ensuring sustainable practices in the wine industry is critical to ensuring continued success and growth within the sector.

As ecologically and environmentally mindful production processes are becoming more important to manufacturers and consumers alike, water stewardship has emerged as an effective means to reduce the environmental impact of the production of numerous goods. A number of drivers and barriers to improved water stewardship specific to the wine-making industry in Italy have been identified by Aivazidou and Tsolakis [2], and are applicable to wine producers worldwide (Table 1).

The Water Footprint (WF) assessment was developed to address a need for a global standard to quantify the impact of human activity on local water resources [28]. This approach addresses one of the critical barriers to effective water stewardship by providing a framework for a standardized approach and methodology in the assessment of water resource impacts. This standardization provides a means to benchmark performance, allowing the direct comparison of a production facility's performance to other similar facilities. It also allows the implementation of mitigation measures to be prioritized to reduce and quantify the reduction in environmental impacts.

The wine-making process can be divided into two main components: viticulture and vinification [25]. The first is an agricultural process, with grapes as the final product. The latter uses fresh grapes, the key ingredient, and a number of mechanical and chemical processing steps to produce wine, the final product. Because of the mix of agricultural and industrial processes, the wine-making utilizes a complex mix

Table 1 Water stewardship drivers and barriers in the wine-making industry (adapted from [2])

Drivers	Barriers
• Sustainable marketing	• Geographic limitations and differences
• Consumer behaviour	• Limited marketing differentiation
• Correlation between wine quality and local water resource health	• Lack of standardization in approach and methodologies
• Production effectiveness	
• Recent benchmarking initiatives	
• Funding and policies for water management	

of rain and freshwater (surface and/or ground) sources, while also generating both non-point and point-source discharges of contaminants to the natural environment. Effluent water reuse for irrigation within the vineyard is becoming more popular as a way to reduce extraction of local surface and/or groundwaters [13, 47], adding further complexity to the evaluation of the overall impact of winery production operations on water resources.

Due to the climate and soil conditions needed to support grape growing, winery operations are usually concentrated in small geographic regions [19]. Some well-known wine regions include Champagne and Bordeaux in France, Rheingau and Rheinhessen in Germany, Napa Valley and Paso Robles in California, and the Niagara Region and Kootenays in the Canadian provinces of Ontario and British Columbia, respectively. Understanding water consumption and wastewater treatment performance is critical to assessing options that can be implemented at a particular facility to conserve water and reduce environmental impacts. By contrast, regional assessments may be appropriate for assessing the overall impact of winery operations on local water resources, since the combined impact from multiple wineries can have significant impacts on certain water resources such as specific aquifers and surface water bodies [27].

A number of methodologies have been proposed and continue to evolve in an effort to identify reliable approaches to quantify the environmental impact associated with the appropriation of freshwater resources [50]. This, coupled with the complexity of the water extraction and discharge cycles during wine-making and the impact of site-specific conditions on water demand and use, have resulted in wide variations in the interpretation and application of the WF concept to wine-making. In some cases, assessments focus on only a portion of the overall WF, while in other cases a complete WF value is reported; some consider only a single winery while other assessments are regional in scope; some consider the footprint from vineyard to bottle, while others consider only a fraction of the processes involved in wine-making. Modifications and enhancements to the approach as first proposed by Hoekstra et al. [28] have been made in recent years, making water footprinting an evolving assessment tool. Some of these proposed enhancements and modifications apply to the WF assessment approach from a Life Cycle Assessment (LCA) perspective, such as ISO 14046:2014 [31], AWARE—Available WAter REmaining [12]—and the Water Stress Index [53], while the V.I.V.A. (Valutazione Impatto Viticoltura sull'Ambiente) tool has been developed specifically for the assessment of the consumptive WF associated with wine-making process [39].

Here, we review the wine-making process and factors that affect water use and pollution; recent advances in water use and conservation; winery wastewater treatment and effluent reuse; and an overview of the approaches available to define the scope of and calculate the WF of wine-making. Recent applications of the WF assessment process to case studies, and challenges associated with utilizing these approaches, are also discussed.

2 Overview of the Wine-Making Process

The process of wine-making can be divided into two main stages: viticulture, the agricultural phase yielding grapes as the final product; and vinification, turning the grapes into wine. Although other fruits can be used in the wine-making process, such as apples, pears, plums, peaches and a variety of berries [38], this chapter focuses on the traditional form of wine-making that is based on grapes alone.

The vineyard is the heart of the viticultural phase of wine-making. Total global agricultural area dedicated to grape growing for wine, table grapes and raisins is estimated at a steady 7.4 million ha [29]. The grapevine species *Vitis vinifera* is the source of almost all of the world's wine-producing grapes. To address the worldwide introduction of a damaging pest, *Phylloxera vastatrix*, the *Vitis vinifera* vine is grafted onto the rootstock of pest-resistant American vine species that produce lower quality fruit. Through grafting, it is possible to choose rootstock best suited to the local climate and soil conditions of the vineyard while maintaining fruit quality. The lifespan of a vine can be up to 100 years, although most are replaced sooner [25]. While older vines are considered to be the source of better quality fruit, the yield decreases as vines age [58].

Vine density varies significantly from vineyard to vineyard, ranging from 1,100 to 10,000 vines per hectare [33]. Factors that influence spacing and density include soil quality, harvesting methods (with mechanical harvesting requiring more spacing between vines than manual harvesting), and, often, tradition. Most New World wineries tend to have lower density vineyards, while more traditional regions, such as Burgundy and Bordeaux, have some of the highest densities [25, 33].

The grape berry is comprised of a number of components. The stalks, rich in tannins, may or may not be completely removed prior to vinification. The skins are rich in other tannins, such as anthocyanins, that add colour, aroma and flavour to wines, and are particularly important in the production of red wines and roses [44]. Wild yeasts and other bacteria are also found on the outer waxy layer of the skin, called the bloom. While wild yeasts are sometimes used in the fermentation process, they generally cannot tolerate ethanol concentrations greater than approximately 4% v/v; therefore, wine yeasts that can tolerate concentrations of up to 15% v/v are added during the fermentation stage of vinification [25]. The pulp of the grape berry contains water, sugars (glucose and fructose), fruit acids (with tartaric acid the most predominant when ripe), proteins and minerals (with potassium the most abundant at up to 2,500 mg/L) [25]. The pips, or seeds, are high in tannins, and can impart a bitterness to the wine if they are split during vinification.

In the northern hemisphere, the grape harvesting is typically between August and October, while spanning the period February to April in the southern hemisphere. The production of ice wine requires sustained temperatures of $-8\,°C$ or lower prior to harvest, and the northern hemisphere wineries that produce this sweet wine typically harvest between December and January, with the exact date dictated by local weather conditions [63]. Because the vine growth stops at temperatures less than $10\,°C$, the winter provides a period of rest for the vine [55]. In some areas that are warm year-round, such as Brazil, it may be possible to obtain up to 5 harvests over a two-year period, however, this will cause the lifespan of the vine to decrease accordingly [25].

The winery is the centre of the vinification stage. The harvested grapes are brought to the winery as quickly as possible for processing, making the vinification stage seasonal in nature. The harvested grapes are destemmed and crushed to produce the must, which is the mixture of grape juice, skins and/or stems that are directed to the fermentation step. In the case of white wines, skins are removed prior to fermentation, either during crushing or after a short contact period with the juice. Because the berry pulp is essentially colourless, the skins of black grapes are left in the must for the production of red wines, since it is the leaching of these coloured compounds from the skins that produce red wine's distinctive colour. Prior to fermentation, the must may undergo several processing steps, including treatment with SO_2, which acts as a disinfectant and antioxidant, as well as acidification or deacidification, as needed. Sucrose may be added, a process called chaptalization, however some regions, such as California, Italy, Greece and Spain, prohibit the addition of sugar to the must [62]. Because of the high water content of the grape berry pulp, water is not generally added as an ingredient during vinification and, like chaptalization, its addition is often prohibited; despite this, it is occasionally permitted if the sugar concentration in the must is very high [59]. In its simplest form, the fermentation step of the vinification stage can be described as follows [25]:

$$Glucose + Fructose + OtherCompounds$$
$$\rightarrow Ethanol + CO_2 + Other Products(small quantities) + Heat$$

Because of the heat generated, the temperature within the fermentation vessel can quickly rise and result in fermentation inhibition at temperatures above 35 °C [25]. For this reason, wineries are typically equipped with temperature control systems to maintain target temperature setpoints throughout the fermentation process. At the conclusion of fermentation, the wine is racked by transferring the liquid from one vessel to another, leaving solids and other sediments behind. These solids can then be pressed, if desired, to obtain more free-run juice. The fermented juice may then undergo malolactic fermentation, during which bacteria convert malic acid to lactic acid. Blending, the process of mixing multiple batches of fermented juice together to produce a blended wine, is also often practiced. Maturation, which may take only a few days to weeks for low-quality wines to up to 22 months for higher quality wines, is then achieved in stainless steel or wooden barrels [18].

Prior to bottling, the remaining fine colloidal solids are removed via fining, a coagulation process that utilizes fining agents such as bentonite, kaolinite, gelatin and isinglass, among others [42]. Filtration is typically used to remove any remaining solids immediately prior to bottling. Filtration techniques commonly used include earth filtration, plate and frame filtration, or membrane filtration. Filtration may also be used during other stages of the vinification stage, such as filtering the free-run juice prior to maturation.

The viticultural and vinification stages may be conducted at a single location, with the winery located directly adjacent to the vineyard. Alternatively, a winery may act as a stand-alone operation, with grapes acquired from offsite vineyards. Similarly, a vineyard may act as a stand-alone operation, shipping grapes to one or more wineries. Some wineries may choose to supplement the grapes grown on-site with additional fruit from another vineyard. No matter where they are grown, a WF was exerted in the production of the grapes.

3 Water Use and Conservation

To develop an accurate estimate of the WF of wine production, it is important to understand water use practices and conservation methods available throughout the process. The following sub-sections outline the way in which water is used during viticulture, vinification and other areas of the vineyard and winery operations.

3.1 Water Usage During Viticulture

The single largest use of water abstracted from the environment during viticulture is for irrigation of the vineyard. Irrigation needs vary geographically, and are strongly affected by local climate and weather conditions. In areas that receive 500–700 mm of rain annually, irrigation may only be required periodically to address a prolonged dry spell [25, 27, 56]. Despite this, wine-making can flourish in drier climates provided irrigation rates can meet the needs of the vines to support grape berry growth, such as the Mendoza area of Argentina, a desert climate which sees an average annual precipitation as of less than 250 mm [46]. Irrigation technologies include furrow, overhead sprinkler and drip (trickle) irrigation [33].

In addition to local climate and weather patterns, the quality and hydraulic properties of soil in the vineyard can impact the total volume of water required to sustain the vines and ensure proper growth of the grape berries [27]. While most of the vine roots are located in the top soil, it is the deeper soil layers into which some roots burrow that are the most important factor affecting availability of water to the vines [25]. Good drainage is also needed for proper vine growth. Chalk, which is porous, promotes the storage of water over the winter while also providing good drainage. Sand, by contrast, provides virtually no storage of water or nutrients, but it is resistant to infestations of *Phylloxera*. Vines can be grown in many different soil types including clay, gravel, granite, limestone and slate, each providing varying levels of water storage and drainage. It is, therefore, possible that irrigation requirements can vary even within the same vineyard if soil conditions vary spatially.

Water utilized by the vines is not only incorporated into the grape berry, and, therefore, directly into the final wine product, it is also used for plant growth. Leaves are also occasionally trimmed from the vines to improve airflow and increase sunlight reaching the berries [17], which also has the effect of reducing transpiration. Despite this, a minimum leaf surface area is required to support grape berry production, estimated to be between 7 and 14 cm^2/g of fruit [25]. Water from both rain and irrigation is also lost to direct evaporation from the soil to the atmosphere. The combined losses due to transpiration and evaporation are often reported as a combined Evapotranspiration (ET) value. The majority of the water loss in the vineyard is lost to ET [27]. Finally, overland losses due to surface runoff can occur during heavy rains [7]. Given the fact that most vineyards are on relatively flat or terraced ground, such overland losses are often minimal [25]. In sloping vineyards, the presence of grass cover reduced runoff rates by as much as 40% [7].

For wineries equipped with an on-site wastewater treatment system, it is sometimes possible to utilize treated effluent as a source of irrigation water, reducing the burden on surface and groundwater resources. The quality of the treated effluent can, however, impact soil quality and grape berry yield and quality. These factors must be taken into account when assessing the volume and location of effluent discharges [13]. This is discussed in more detail in Sect. 4.4.

Finally, water can be used to dilute full-strength industrial chemicals, such as pesticides and herbicides, prior to their application in the vineyard. Water aspersion systems can also be used for frost protection in the spring season. By spraying and coating buds with water prior to an anticipated frost while temperatures are still above freezing, the resulting ice layer provides protection to the buds, which can mitigate frost damage [61].

3.2 Water Usage During Vinification

The water in the grape berry pulp provides the liquid in the grape must needed for wine production. This differentiates wine from other types of beverages, such as beer and soft drinks, that require substantial volumes of water as direct ingredients in the production process, representing up to 70% of the total water use in the bottling plant [20]. Water is not generally used as an ingredient in the wine-making process unless dilution of the grape must is needed to reduce high sugar concentrations [59]. While some jurisdictions, such as Australia, allow the addition of water to grape must under certain circumstances [5], others, including most of Europe, have banned the practice [58].

Despite this, large volumes of water are used during vinification for equipment and facility cleaning and sanitizing. Water use estimates range from approximately 1 L/L of wine to 10 L/L of wine [16, 25]. Larger wineries tend to have lower water use rates, with values reported by Storm [56] ranging from as low as 4.2 L/L for wineries with a production capacity of more than 1,000,000 cases per year, to as high as 10.6 L/L of wine for those with a capacity of less than 50,000 cases per year.

Annual wine production volumes can also impact normalized water use values (L water/L wine produced). A benchmarking study conducted by the Beverage Industry Environmental Roundtable [6] reviewed water use at 27 wineries over a 3-year period and found values ranging from 2 to 18.5 L/L, and that the variability in annual production volume (L wine/yr) had a considerable impact on the normalized water use value. Therefore, total water use trends (L water use/yr) and fluctuations in overall production (L wine/yr) should also be considered when evaluating a winery's water use data.

The techniques used for sanitizing equipment can have a significant impact on water use, which contributes to its wide variability. For example, Garcia-Alcaraz et al. [23] compared the water use associated with four wine barrel sanitizing techniques, and found that a water vapour + SO_2 technique required over three times the volume of water than a CO_2 disinfection technique.

Overall water consumption can be reduced via the adoption of industry best practices. The water audit approach [32, 57] can be used to identify and prioritize water conservation opportunities. In addition to utilizing low-consumption cleaning and sanitation techniques, options can include installing flow metres to allow monitoring of water use, proactive squeegeeing to remove wastes and spillage prior to washing, and implementing spill prevention measures [16].

3.3 Other Areas of Water Usage

Many wineries now take part in the "wine tourism" sector to increase revenues [37], but with these come additional water demands. In addition to requiring the provision of washroom facilities for tourists, some wineries may also have on-site restaurants and accommodations. Irrigation of lawns and shrubbery may also be implemented to increase visual appeal of the property. Despite the water use associated with these activities, these increased demands do not directly impact the WF of the wine-making process, since they are not directly related to the manufacturing of the wine. By contrast, the water demand exerted by vineyard and winery employees, such as in washroom and lunchroom facilities, directly affect the WF of the wine-making process since these employees are directly involved in the viticultural and vinification stages of production.

Although offsite, water is used in the manufacturing of the chemicals, cleaning agents, bottles, labels, glue and corking materials that are ultimately used by the vineyard and winery. When available, the data for these water uses can be included in a water use assessment, although it should be recognized that these are not under the control of the vineyard or winery operations. If included in a WF assessment, this water use would contribute to the indirect WF (see Sect. 5.1).

4   Wastewater Generation, Handling and Fate

In addition to water use, an understanding of wastewater handling, treatment and disposal is required to develop a comprehensive understanding of the WF of the wine-making process. While effluents from wastewater holding or treatment systems represent point-source discharges of contaminants entering the natural environment, non-point sources of pollution to natural water bodies are an important consideration when assessing the environmental impact of wine-making on local water resources. This section reviews the main sources of point and non-point-source contaminant discharges, as well as an overview of current winery wastewater handling, treatment and reuse strategies used at full scale.

4.1 Wastewater During Viticulture

The main source of wastewater in viticulture is drainage and runoff from the vineyard, which can contaminate both surface and groundwaters as a non-point source of pollution. Typical compounds of concern include nutrients, such as nitrogen and phosphorus, and chemicals applied for pest control, including pesticides and herbicides. Pathways for contamination include runoff to surface body waters, or drainage through the overburden into groundwaters. In addition to pesticides and nutrients, sediments can also be found in vineyard runoff [7], which can contribute to heavy metal and turbidity pollution of local waterways. Pesticides and herbicides applied to the vineyard can also be prone to drifting, which can result in their ultimate deposition into and pollution of nearby surface waters [25].

Due to their nature, pesticides and herbicides can have toxic impacts at very low concentrations. Reducing their discharge to the environment can substantially reduce the environmental impacts of vineyard operations. Recent advances to reduce loadings to the environment include the development of natural, biological approaches to pest control. Planting hedgerows at the edges of the vineyard are effective at controlling drift, with a mature hedgerow able to reduce drifting by up to 75% [39]. In addition, the application of some pesticides can be reduced or eliminated by flowering plants that attract the predators of pests, while other options include allowing higher order predators, such as ducks and geese, access to the vineyard [25]. Such biological approaches can, however, take 4–5 years to become fully effective.

Physical barriers applied to the vine trunks can also prevent insects from reaching the leaves and fruit, reducing the need to apply pesticides. When treatment is needed, more precise pesticide application methods (such as directed spraying) and spray techniques can also reduce drifting. Reductions in drifting of up to 50–90% are possible when using precision instruments or tunnel sprayers, respectively, versus conventional nozzles [39]. Better weather forecasting also allows more precise timing of application, which can be particularly useful to reducing the application rates of chemicals used for mildew control such as sulphur, copper sulphate and lime [25].

Nutrients in agricultural runoff are a concern for the health of local surface waters. Two key nutrients required for plant growth, phosphorus and nitrogen, can promote eutrophication and are often added as fertilizers to crop land. Phosphorus fertilizers are used mainly for younger vines, with older vines requiring less phosphorus [61], reducing long-term fertilization needs. Legumes planted as cover crops also provide nitrogen fixing, further reducing fertilizer needs [25]. Physical runoff can be reduced by planting perennial cover crops (up to 50% reduction) or inter-row soil working (up to 40% reduction) [39].

4.2 Wastewater During Vinification

During the vinification stage, cleaning tanks, crushers, presses/filters, piping and bottles are the main source of wastewater. Being for human consumption, the production process must meet high cleaning and sanitation standards, requiring the use of sanitizing chemicals and large volumes of water. Other significant sources of wastewater include spillage during transfer and/or overfilling vessels which can be reduced by implementing better process control strategies to minimize transfers and active level control [16]. While the rates of Winery Wastewater (WWW) generation during vinification vary from winery to winery, and are driven by the volume of water used for cleaning and sanitation, the proportions of wastewater that can be attributed to each production step have been estimated, and are shown in Fig. 1.

While the values shown in Fig. 1 can be used for the purposes of high-level estimations, the actual proportions will vary from winery to winery based on-site-specific conditions. For example, the type(s) of material(s) used for tanks and vessels can significantly impact cleaning water needs. Stainless steel maturation vessels are the easiest to clean, while wood barrels, due to their porous nature, must generally undergo more complex and involved cleaning and sanitization activities [18]. Despite this, wood barrels may be preferred for certain wineries and/or specific products as volatile compounds present in the wood add colour and aroma to both red and white wines [3].

Due to the wide range of processes generating WWW during vinification, parameter concentrations can vary by up to four orders of magnitude. The highest strengths are generally associated with the vintage period, when winery activities include

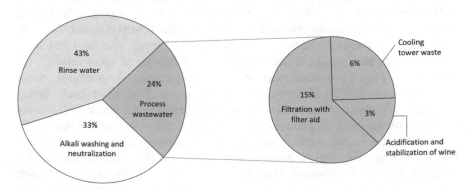

Fig. 1 Fraction of winery wastewater associated with various vinification processes (adapted from [16])

crushing, fermenting and racking; lower strengths are typical during bottling. Table 2 presents reported variability in WWW characteristics for select parameters.

Key parameters of concern include Biochemical Oxygen Demand (BOD) and Chemical Oxygen Demand (COD) which can result in low DO concentrations in receiving streams. Na^+ is often found in high concentrations due to sodium-based cleaning agents used for sanitation, while K^+ is present in high concentrations in the grape berry pulp [25]. Ions are not removed via typical wastewater treatment techniques [15], therefore, high Na^+ and K^+ concentrations can reduce effluent reuse potential (see Sect. 4.4).

Process wastewaters are generally the highest in strength and lowest in pH. The first flush of tank rinsing water can be very high in COD and other organics, while subsequent flushes are lower in strength. Cleaning water is generally high in pH as well as Na^+ or K^+, depending on the specific cleaning chemicals used [40]. Bottle washing and disinfection prior to bottling is generally the lowest strength and highest in pH [13, 16].

Spoiled product, which is of very high strength particularly with respect to soluble parameter fractions, can also be a source of wastewater requiring treatment and disposal [35]. Lees from racking are also very high strength and, due to their nature, are high in solids and particulate parameter fractions. These very high-strength wastewaters can overwhelm on-site treatment systems and, therefore, it is recommended best practice to keep these separate from other wastewaters, if possible, for subsequent handling and treatment [16]. For example, in the Niagara Region of Ontario, Canada, wineries equipped with on-site treatment systems will often direct very high-strength wastewaters to local municipal Wastewater Treatment Plants (WWTPs) rather than treating on-site [35].

Despite the wide variability in WWW characteristics, relationships between various parameters have been successfully developed [14, 36]. Rapid, reliable and cost-effective methods to characterize WWW can improve performance of WWW treatment systems and inform decisions with respect to effluent reuse for irrigation that will not negatively affect soil characteristics [13, 14].

On-site processes typically used to treat the WWW include simple lagoon-based or constructed wetland systems [22, 35, 43]. These systems can be used to treat WWW alone, or in combination with sanitary sewage generated on-site. Despite the flexibility of on-site treatment systems, they can be subject to overloading due to the high-volume and high-strength discharges typical during the vintage period. To address these operational limitations, some wineries will haul some or all of the WWW during these periods to other locations for treatment, with local municipal WWTPs being the most common [41]. This can, however, have negative impacts on the performance of the municipal WWTP and negatively affect effluent quality [8, 9, 35]. In addition, treating a portion of the WWW offsite results in multiple point-source discharges of treated WWW that need to be considered when assessing the overall environmental impact of WWW effluents. The on-site wastewater treatment systems may themselves produce waste streams that also require treatment. Constructed wetland systems, for example, are often equipped with simple settling tanks at the head of the

Table 2 Reported variability of winery wastewater characteristics

Parameter	Unit	Range of values reported in the literature			References
Conductivity	μS/cm	80	–	9700	[1, 4–7]
SAR	mEq/L	0.21		46.9	[1]
pH	–	2.5	–	12.9	[1–7]
BOD$_5$	mg/L	4	–	130,000	[1–7]
COD	mg/L	37	–	360,000	[2–7]
TS	mg/L	190	–	188,000	[2–7]
VS	mg/L	300	–	128,000	[2, 4, 5, 7]
TSS	mg/L	7	–	84,400	[2–7]
TKN	mg/L	0.51	–	14,300	[5]
TAN	mg/L	0.20	–	395	[5]
TP	mg/L	0.70	–	1120	[4, 5]
Al	mg/L	0.04	–	1030	[5]
Ca	mg/L	1.8	–	2203	[2, 5]
Cl$^-$	mg/L	3.4	–	2290	[1, 5]
Fe	mg/L	0.17	–	335	[2, 5]
K	mg/L	1.3	–	8000	[2, 5]
Mg	mg/L	1.1	–	530	[2, 5]
Na	mg/L	1.0	–	1160	[2, 5]
Ni	mg/L	0.003	–	3.0	[2, 5]
Zn	mg/L	0.012	–	46.0	[2, 5, 6]
Cu	μg/L	3	–	29,600	[2, 5, 6]
Hg	μg/L	20	–	12,600	[5]
Mn	μg/L	100	–	61,000	[2, 5]
Pb	μg/L	5.0	–	2920	[2, 5]
Ni	μg/L	3.0	–	3000	[2, 5, 6]
Ethanol	mg/L	1000		5000	[3, 7]
Glucose	mg/L	0	–	2700	[7]
Fructose	mg/L	0	–	1500	[7]
Tartaric acid	mg/L	0	–	1260	[3, 7]
Malic acid	mg/L	0	–	70	[3, 7]
Lactic acid	mg/L	0	–	350	[3, 7]
Total phenolic compounds	mg/L	0.51	–	1450	[4, 6]

Notes Reference numbers refer to the following: 1. [13]; 2. [14]; 3. [16]; 4. [30]; 5. [36]; 6. [41]; 7. [49]

process [35, 43, 47] that require pump down to remove collected sediment. Lagoon-based systems require periodic dredging to remove accumulated sludge in order to maintain adequate operating volume and treatment performance.

Where proximity allows, wineries can be connected directly to the sanitary sewer system. Pre-treatment of WWW prior to discharge and/or surcharge fees may apply, depending on local bylaws or other legislative requirements. The WWW is then conveyed to the local municipal WWTP for treatment along with other domestic, institutional, commercial and industrial wastewaters collected in the sewershed.

The vinification stage also results in the generation of a solid waste, called grape marc or grape pomace, which is composed of grape skins, seeds, stems, etc. The rate of grape marc generation is estimated to be 0.17 kg/L of wine [24]. This solid waste can be used for the production of grape spirits [51], which are more commonly known by regional names such as grappa (Italy) and marc (France). Grape pomace can be used directly as an additive in animal feed, or treated using aerobic or anaerobic biological digestion with the resulting digestate or compost utilized as a soil amendment [24]. Improper handling, treatment or disposal of the solid grape marc waste could result in contamination of surface or groundwaters. Despite this, use of grape marc for environmental remediation, particularly for heavy metal removal and recovery from industrial effluent, has shown promise [51].

4.3 Wastewater in Other Areas

As noted in Sect. 3.3, incorporating "wine tourism" activities at wineries increases water demands and, as a result, also increases wastewater generation rates from facilities including restrooms, restaurants and accommodations, among others. As with water demands, only the wastewater generated by employees directly involved in the viticultural and vinification stages of production would contribute to the direct WF of the wine-making process.

Additionally, wastewater is generated during the manufacturing of the chemicals, cleaning agents, bottles, labels, glue and corking materials that are ultimately used by the winery. As with the water demands associated with these products manufactured offsite, these wastewater generation data can be included in an environmental assessment of wastewater effluent impacts. If included in a WF assessment, these impacts would contribute to the indirect WF (see Sect. 5.1).

4.4 Effluent Disposal and Reuse

The discussion that follows is related to the disposal and reuse of treated WWW effluents generated during vinification, but does not include the discharge of non-point-source wastewaters generated during viticulture. Methods to reduce the pollution of water resources during viticulture were discussed in Sect. 4.1.

Effluent disposal can be divided into three main categories: direct discharge to surface waters, subsurface discharge and irrigation discharge. A single winery may use one or more disposal methods. If the wastewater is conveyed to an offsite treatment system, such as a municipal WWTP, then the ultimate disposal method is that used by the offsite treatment system. In such cases, the treated WWW effluents represent only a fraction of the total effluent discharged by the municipal treatment system.

Discharge to surface waters and subsurface discharge, typically via a leaching bed, represent a direct return of treated effluent to the natural environment. Irrigation discharge, where treated effluent is used to supplement or meet vineyard irrigation needs, is the most common reuse strategy used at wineries.

Due to the seasonal nature of viticulture and vinification, peak WWW generation rates do not coincide with peak irrigation needs [56]; therefore, WWW effluents that are reused for irrigation are typically stored in large earthen lagoons that provide storage during the non-irrigation period [35]. These can be part of the treatment process (providing biological treatment), or simple effluent storage ponds. Due to public health safety concerns, effluent reuse for irrigation is generally restricted to winery process wastewater and not sanitary sewage [56], although irrigation reuse of combined process and sanitary wastewater is sometimes possible if adequate disinfection is provided [35].

Utilizing WWW for irrigation can have negative impacts on soil conditions that can affect soil quality and increase the potential for runoff. In particular, the build-up of Na^+ and K^+ ions, which are often present in large concentrations in WWW, can fill void spaces in the soil thus changing its hydraulic conductivity [33]. Typical full-scale WWW treatment methods do not remove ions and, as such, cannot eliminate this potential negative impact [13, 15]. In addition, the application of WWW effluents for irrigation has been found to change the types of microorganisms present in the soil [49].

The ability to reuse effluent for irrigation and the development of maximum application rates (generally expressed as mm/yr) can only be determined through site-specific assessments [13]. These would need to consider the particular soil type(s) present, prior application of WWW effluents, and the characteristics of the vines, particularly rootstocks. Despite these potential limitations, evaluation of treated WWW effluent quality from 18 wineries in California concluded that treatment methods, including physicochemical and biological, were able to produce effluent of high enough quality for irrigation reuse purposes [13]. Other reported WWW effluent reuse for irrigation includes jurisdictions in Italy [47], South Africa [52] and Canada [35].

5 Water Footprint Calculations

5.1 Overview and Definitions

The WF of a product is comprised of three distinct components [28]: the rainwater utilized in its production (green WF); freshwater abstracted from the environment during production (blue WF) and the volume of freshwater needed to reduce the concentration of pollutants discharged to water resources during production (grey WF). The total WF can then be defined as the sum of the three WF components as follows:

$$WF = WF_{green} + WF_{blue} + WF_{grey} \tag{1}$$

where the units for WF can be expressed in terms of total volume (m^3), volume per unit time (m^3/d or m^3/yr) or volume per functional unit (m^3/U$_F$). Common functional units used when assessing the WF of the viticultural and/or vinification stages wine-making include kg of grapes, L of wine and 750 mL bottle of wine.

The green and blue components of the WF are direct measurements of water resources abstracted and used during the production process. To calculate these values, two classes of methodologies are reviewed herein, namely:

– consumptive water use, and
– hydrological water-balance.

By contrast, the grey component of the WF is not a direct measurement of water abstracted or used; rather it is the virtual volume of freshwater needed to assimilate loadings of pollutants discharged to water resources during the production process. As such, it cannot be calculated using a direct consumption or water-balance approach, but rather via an assessment that considers the volumes and concentrations of parameters discharged, and the unique characteristics of the receiving stream. Therefore, the calculated grey WF of an effluent discharge can vary significantly depending on the characteristics of the receiver, making the comparison of grey WFs from facility to facility difficult [60].

The WF concept can be further divided into "direct" and "indirect" contributions. The direct WF refers to the WF of processes directly related to the production of a product, while the indirect WF includes the WF inherent in chemicals and materials that may be used at points in the production process [28]. In the case of the wine-making process, items that could be included in the direct and indirect WF assessments of both the viticultural and vinification stages of wine-making are summarized in Table 3.

Various water uses and losses that affect the direct WF of the wine-making process, from viticulture to vinification, are presented graphically in Fig. 2. The factors that affect the direct blue, green and grey WF, as they relate to the wine-making process, are presented in more detail in the following sub-sections. Methods available to calculate the various water footprint fractions of the wine-making process are

Table 3 Examples of factors contributing to direct and indirect WF of wine-making process

Direct	Indirect
Viticulture	
• Evapotranspiration by grapevines	• WF exerted in the production of:
• Evapotranspiration by cover crops	– Pesticides, herbicides, etc.
• Water incorporated into the grape berries	– Fuel used by motorized equipment
• Water used to dilute chemicals	
• Runoff and infiltration of contaminants	
• Drifting of chemicals applied to the vineyard	
Vinification	
• Water used for cleaning and sanitizing	• WF exerted in the production of:
• Water used for dilution of must (if applicable)	– Bottles and other packaging materials
• Disposal of wastewater effluents	– Labels, inks and glues
	– Corking materials
	– Cleaning chemicals

also presented. Because the various units can be used to report water footprints, all equations are presented in terms of general units (L = length, M = mass, T = time).

5.2   Green Water Footprint

Rainwater is not used directly in winery operations, and as such does not contribute to the water footprint of the vinification stage. By contrast, rainwater is used extensively during the viticultural phase. The consumptive green WF approach quantifies the amount of rainwater used by all crops in the vineyard, including cover crops, that is lost to ET. As a result, the consumptive green WF is always a positive value. By contrast, the water-balance approach estimates the net consumption of water stores in the soil. As a result, the day-to-day calculation of the water-balance green WF may be positive (on a day with no precipitation) or negative (on a day with precipitation). Due to this, the water-balance approach yields a negligible overall green WF [27].

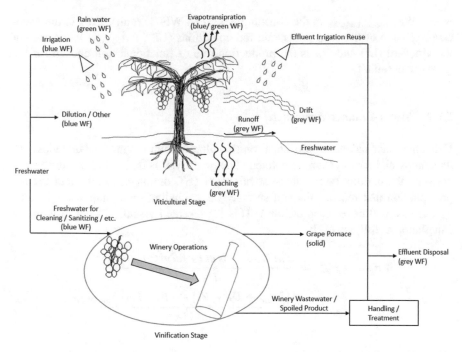

Fig. 2 Simplified graphical representation of water uses and discharges that contribute to the direct water footprint of the wine-making process

5.2.1 Consumptive Approach

Using the consumptive approach, the cultivation of grapes during viticulture is found to exert a large green WF due to the ET of rainwater within the vineyard. The magnitude of the green WF is strongly influenced by local conditions including total precipitation during the growing season, local climatic conditions (temperature, humidity, sun exposure), and soil type. The vineyard configuration, including vine density and presence of cover crops, also influences the green WF. It is possible to calculate the ET for a given vineyard as follows [39]:

$$Et_c = K_c K_s Et_o \tag{2}$$

where Etc is the evapotranspiration of the crop (LT^{-1}), K_c is the crop coefficient, K_s is a soil water stress coefficient considered when moisture depletion exceeds the readily available moisture in the soil and Et_o is the reference crop coefficient (LT^{-1}) as determined by the using the Penman–Monteith equation [4]. Since the consumptive green WF is a measure of the total volume lost to ET, it can then be calculated as follows:

$$WF_{green,consumptive} = Et_{c,r} A_{vineyard} / \dot{U}_F \tag{3}$$

where $WF_{green,consumptive}$ is the consumptive green WF (L³/unit), $Etc_{,r}$ is the evapotranspiration of the crop under rain-fed conditions (LT⁻¹), $A_{vineyard}$ is the area of the vineyard (L²) and $\dot{U}_F$ is the production rate of functional unit chosen for the assessment (unit/T).

5.2.2 Water-Balance Approach

Using the water-balance approach, a control volume around a vine, that includes both its canopy and root system, is defined. The green WF is defined as the difference between the outflow components of rainwater (ET, drainage, runoff) and the net precipitation that reaches the soil surface (i.e. excluding precipitation intercepted by the vine and cover crop canopy). This is presented visually in Fig. 3 with the calculation as follows [27]:

$$
WF_{green,balance} = \frac{Outflow - Net\,Precipitation}{\dot{U}_F}
$$

$$
= \frac{\left[\left(Et_{c,r} + D_r + R_r\right) - \left(P - P_i\right)\right]A_{vineyard}}{\dot{U}_F} \tag{4}
$$

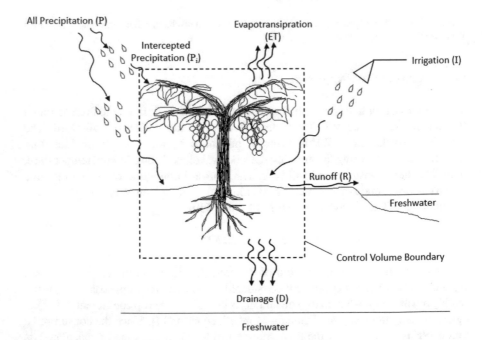

Fig. 3 Control volume used for the water-balance footprint approach for the viticultural stage of wine making (adapted from [27])

where $WF_{green,balance}$ is the water-balanced green WF (L^3/unit), D_r is the drainage back to a groundwater resource under rain-fed conditions (LT^{-1}), R_r is the surface runoff under rain-fed conditions (LT^{-1}), P is the precipitation (LT^{-1}), P_i is the precipitation intercepted by the canopy that does not reach the ground (LT^{-1}).

5.2.3 Discussion

The green WF during viticulture applies to the use of rainwater during the growing season, and should, therefore, neglect the WF impact of irrigation using freshwater sources since this contributes to the blue WF (see Sect. 5.3). As a result, parameter values needed for the calculation of the green WF (such as ET, drainage and runoff) should be developed under rain-fed conditions. In Eqs. (3) and (4), this is denoted by the subscript r. In addition, Eqs. (3) and (4) do not include a term for the rainwater that is accumulated within the grape berry. Based on published WF assessments [27, 39], the water stored in the grape berries would only represent 0.10–0.15% of the rainwater lost to ET. While this volume of rainwater would contribute to the green WF, its magnitude is negligible compared to that lost to ET and, as such, can be neglected.

5.3 Blue Water Footprint

The blue WF is a measure of freshwater abstracted from a natural freshwater source. In contrast to the green WF, both the viticultural and vinification stages of wine-making exert a blue WF. The blue WF does not take into consideration any pollutants or contaminants that are discharged when the freshwater is returned to the natural environment; rather, these are considered when evaluating the grey WF (Sect. 5.4).

5.3.1 Consumptive Approach

Using the consumptive WF assessment approach, the total blue water footprint is the sum of the volumes of freshwater abstracted from the environment that are incorporated into a product; lost to a evaporation/ET; used and discharged to a different source from which it was taken; and/or used and returned to the same source after being held for a prolonged period of time.

During viticulture, the blue water footprint is associated with freshwater used for irrigation that is lost to ET, as well as any water used for diluting chemical products and frost control.

During vinification, the blue WF represents the volume of freshwater used in the production process that is not returned to the source from which it was taken. Most of the freshwater use during this stage is associated with cleaning activities, while very little is incorporated into the product or lost to evaporation. Therefore, provided the freshwater used during the vinification step is returned to the source from which it was taken with minimal delay, the blue WF of the vinification step could be negligible. However, in practice, the water used for cleaning and sanitation generates WWW that requires treatment before it can be returned to the environment. Because there is often a delay in the treatment and discharge of effluents and/or the treated effluent is not always returned to the freshwater source from which it was abstracted, the blue WF of vinification is generally reported as the volume of water used for winery operations. The impact of any pollutant loadings associated with effluent discharge would be addressed by the grey WF. The blue WF determined via the consumptive approach can, therefore, be defined as follows:

$$WF_{blue,consumptive} = \left[Et_{c,i} A_{vineyard} + Q_{vineyard} + Q_{winery} \right] / \dot{U}_F \qquad (5)$$

where $WF_{blue,consumptive}$ is the consumptive blue WF (L^3/unit), $Et_{c,i}$ is the ET under irrigated conditions (LT^{-1}), $Q_{vineyard}$ is the volume of water used for diluting chemicals, frost control or other vineyard activities excluding irrigation (L^{3}T^{-1}) and Q_{winery} is the volume of water used for winery operations (L^{3}T^{-1}). Equation (2) can be used to calculate $Et_{c,i}$.

5.3.2 Water-Balance Approach

Using the water-balance approach, the calculation of the blue WF during the vinification step is the same as that used in the consumptive approach, and can be generalized as the amount of water used in the winery for cleaning, sanitation and other activities per functional unit (Q_{winery}/U_F). During viticulture, however, the blue WF calculated using the water-balance approach represents net water resource usage associated with vineyard operations. In contrast to the consumptive approach, the water-balance blue WF can yield positive or negative numbers. Using the approach outlined in Herath et al. [27], the blue WF can then be defined as follows:

$$WF_{blue,balance} = \left\{ [I - (D + R)] A_{vineyard} + Q_{vineyard} + Q_{winery} \right\} / \dot{U}_F \qquad (6)$$

where $WF_{blue,balance}$ is the water-balanced blue WF (L^3/unit), I is the freshwater abstracted for irrigation purposes (LT^{-1}), D is root zone drainage during the growing season (LT^{-1}) and R is vineyard runoff during the growing season (LT^{-1}).

5.3.3 Discussion

As noted in Sect. 5.3.2, the water-balance approach yields a negligible green WF; thus, it is the blue WF that gives an overall snapshot of the impact of wine making on local water resources. If the water-balance blue WF is positive, then this indicates that there is a depletion of local water resources (i.e. a net loss from surface and/or groundwaters); if it is negative, then the freshwater lost to the wine-making processes does not deplete the natural water source(s) (i.e. a net increase to surface and/or groundwaters when accounting for both water abstracted and natural recharge rates). By contrast, the consumptive approach provides information regarding the volume of rain and freshwater that is lost to ET during viticulture and/or is not returned to the source from which it was abstracted during vinification.

As noted in Sect. 4.4, WWW effluents are often used for irrigation. Such reuse activities would reduce the consumptive blue WF by reducing the volume of freshwater that needs to be abstracted from the environment for irrigation purposes. As water is abstracted for process needs, this is captured in the Q_{winery} term of Eqs. (5) and (6). Under such reuse conditions, the values of Etc_i and I from Eqs. (5) and (6), respectively, would need to be adjusted to ensure that irrigation via effluent reuse volume is not included to avoid double-counting this volume in the blue WF calculation.

5.4 Grey Water Footprint

5.4.1 Approach

Rather than representing a direct measurement of the amount of water that is abstracted from the environment, the grey WF represents the virtual volume of water already present in the environment that is needed to dilute freshwater pollution during wine making. Using the definition from Hoekstra et al. [28], the grey WF during viticulture would be driven by the volume of freshwater required to dilute contaminants from vineyard leaching, runoff and drifting.

During vinification, the grey WF would be equivalent to the volume of freshwater required to dilute pollutant loadings from the discharge of WWW effluents into the natural environment. If effluent is used for irrigation purposes, the grey WF associated with any potential runoff or drainage could be captured in the assessment of viticultural grey WF and should, therefore, be excluded from the grey WF of vinification. If WWW effluents are discharged to the environment via multiple point-source discharge locations (e.g. treatment of a portion of the WWW via an on-site treatment system with another fraction treated offsite, such as at a municipal WWTP), then the overall grey WF would be the sum of the grey WFs calculated for each discharge location.

Regardless of the wine-making stage or processes under consideration, the grey WF is calculated as the minimum volume of water required to dilute the applied contaminant loadings to the maximum allowable concentration in the receiver [28], as follows:

$$W F_{grey} = \frac{\dot{m}_e - \dot{m}_a}{C_{max} - C_{nat}} \tag{6}$$

where WF_{grey} is the grey water footprint ($L^3 T^{-1}$), $\dot{m}_e$ is the mass loading rate of a parameter discharged to a receiver (MT^{-1}), $\dot{m}_a$ is the mass loading rate of a parameter abstracted from the receiver that is incorporated into the product and effluent stream (MT^{-1}), C_{max} is the receiver target maximum concentration (ML^{-3}) and C_{nat} is the natural concentration in the receiver (ML^{-3}).

During viticulture, loadings to the environment ($\dot{m}_e$) would be those associated with leaching, runoff and drifting to a particular receiver. Lamastra et al. [39] present techniques to calculate these values. Loadings removed from the environment ($\dot{m}_a$) would be the loadings associated with water taken from the receiver and used for irrigation or other vineyard purposes (diluting chemicals, frost protection, etc.). During vinification, $\dot{m}_e$ would represent the loadings from wastewater effluents discharged to a natural water body after treatment, if any, while $\dot{m}_a$ would represent the loadings of parameters in water abstracted from the receiver for process purposes (cleaning, sanitizing, etc.). During vinification, the parameter concentrations in treated WWW effluents are often orders of magnitude greater than that present in the receiver. In addition, water is often abstracted from one source (such as a well) while discharged as WWW effluent into another (such as a river or lake). Under such circumstances, it can be assumed that $\dot{m}_e << \dot{m}_a$ and Eq. (6) can be simplified to the following equation [34]:

$$WF_{grey} = \frac{\dot{m}_e}{C_{max} - C_{nat}} \tag{7}$$

5.4.2 Discussion

Contamination of the environment during both viticulture and vinification is due to the discharge of more than one pollutant in each waste stream. The grey WF is defined by the critical parameter, which is the pollutant that results in the largest calculated footprint. Therefore, a grey WF analysis requires the calculation of the footprint associated with a number of parameters present in each waste stream. This requires defining the appropriate loading and concentration values, as specified in Eq. (6) or Eq. (7), for each parameter. Furthermore, if more than one receiver is polluted (e.g. groundwater aquifer subject to contamination from leaching while a nearby river is polluted due to runoff and/or drifting), the grey WF should be calculated for all affected receivers and summed to determine the overall grey WF.

Selection of appropriate C_{max} and C_{nat} values is critical for the development of accurate grey WF estimates. Regulatory bodies, where applicable, can be used to define the maximum allowable concentration (C_{max}) in a given receiver based on-site-specific conditions and targets. Defining an appropriate value for C_{nat} is, however, more complex: this concentration is defined as the expected concentration in a receiver if it were not subject to human interference [28]. As a result, C_{nat} cannot be measured directly. Two approaches have been used to select an appropriate value for C_{nat}: the first assumes that the observed background water quality in the receiver provides a reasonable estimate of C_{nat} [48]; the second assumes C_{nat} is equal to zero [28]. The grey WF is overestimated by the first approach, and underestimated by the second.

The proximity and known impacts of human activity on background water quality, as well as the magnitude of background concentrations as they compare to C_{max}, should be considered when selecting an appropriate value for C_{nat}. In addition, there may be spatial variations in the characteristics of a receiver that affect background concentrations (such as confluences with other water bodies or discharges of other effluent streams) or maximum allowable concentrations (such as presence of species at risk or drinking water intake zones). Therefore, careful evaluation of the actual discharge location and site-specific characteristics of the receiver should be incorporated into a grey WF assessment, as these factors can significantly affect the magnitude of the calculated footprint [34, 60].

Finally, two approaches have been used to estimate the grey WF related to the discharge of WWW effluents to the environment, with significant variability in the magnitude of the calculated footprint associated with each. The first approach, which results in a grey WF less than the volume of the effluent discharged, assumes that the effluent meets all regulatory limits applicable to the receiver, or that the effluent concentrations for all parameters are less than the corresponding value of C_{max} [10, 21, 27]. The second approach recognizes that wastewater treatment facilities are often designed to meet effluent target concentrates in higher than the receiver water quality target concentrations (C_{max}), resulting in a grey WF that is larger than the volume of effluent discharged despite treatment performance meeting or exceeding treatment targets [26, 34, 48].

5.5 Considerations and Limitations

Selecting the overall scope of the assessment is the first critical step of the WF process. Questions to consider include: Will the WF be developed for a single winery? Will the impacts of only some of the process(es) be considered? Will a regional approach be used that considers all wineries in a geographic area? Limiting the system boundary to a single winery or a subset of the processes in use at a single winery can provide better resolution and more accurate WF estimates. Conversely, a regional assessment is better suited to quantify overall impacts of wine-making activities on local water resources.

Once the facilities and/or processes have been selected, the system boundary can then be defined. While factors affecting the direct WF should be included, the decision to include or exclude factors affecting the indirect WF must also be made. If the purpose of the WF assessment is to reduce the ecological impacts of winery operations on water resources, choosing to exclude the indirect WF associated with the production of bottles, labels, glues, chemicals, etc., may be preferred since the winery cannot control the WF embedded in these materials. Conversely, if the objective is to reduce the overall WF of a product, including these indirect factors may allow decisions to be made to source these materials from alternative suppliers who have implemented their own in-house water stewardship initiatives to reduce their production-related WF. One special case that must also be considered for wine-making involves wineries that do not grow their own grapes and/or supplement the grapes from their vineyard with those grown elsewhere. Because the grapes are a direct ingredient in the wine-making process, the WF associated with their production should be considered a factor contributing to the direct WF of the wine produced, no matter where they are grown.

Where possible, WF assessments should be based on multiple years' worth of data to develop a more accurate estimate. A multi-year approach can smooth out annual variations in non-controllable factors, particularly local weather patterns during the growing season and their effects on irrigation and pest-control needs in the vineyard as well as overall grape berry yield. It is also often advantageous to assess the WF in terms of both a functional unit of product (L/U_F) as well as per unit time (e.g. L/yr). For example, Saraiva et al. [54] reported significant year-to-year increases in winery WF reported in terms of L/U_F (+33% and +32% for blue and grey WF, respectively), despite small reductions in both water consumption (−3%) and pollutant loading (−7%). Inspection of the available data indicated that a significant decrease in year-to-year wine production (−30%) was the cause of the increase in the WF as reported in L/U_F rather than changes in water usage.

The WF concept can be used as a benchmarking tool to compare the performance of subject facility(ies) to other comparable facilities [28]. However, direct "apples to apples" comparisons can only be made when factors specific to the winery(ies) in question (climate, soil type, size) and WF assessment approach (extent of the system boundary, methodology used, WF components (green, blue, grey) included) are also comparable. For example, the performance of a vineyard in an area of Argentina that sees <250 mm/yr of rain cannot be assessed against the viticultural stage green and blue WFs of a vineyard in New Zealand that receives over 1,000 mm/yr. Due to the variability in conditions from winery to winery, selecting facilities that can be used for comparison purposes is difficult, especially given the numerous factors that can affect overall WF. Some significant site-specific factors that should be considered prior to comparing the WF of a winery against benchmark facilities are presented in Table 4.

In terms of methodology, the water-balance approach for green and blue WFs might be best suited to regional assessments, since the operation of the wineries would have a combined impact on water resources in a geographic area [27, 28]. The consumptive approach would be better suited for benchmarking comparisons between wineries and/or developing a performance baseline to prioritize and quantify improvements achieved as operational and/or capital upgrades are implemented.

As noted above, the green and blue WFs provide a quantifiable measure of the amount of water that is abstracted from the environment during the production of a product. Conversely, the grey WF is more conceptual in nature: it is a measure of the virtual volume of water already present in the environment that must be used to

Table 4 Site-specific factors to consider when using reported water footprints as performance benchmarks

Vineyard	Winery
• Climate (temperature, precipitation)	• Type(s) of wine produced
• Soil type	• Annual production output (L wine/year)
• Growing season duration/no. of harvests/year	• Duration of maturation
• Vine density	• Types of tanks/barrels (stainless steel, wood)
• Annual grape production (tonnes/year)	• Sanitizing techniques employed
• Irrigation type	• Sanitizing chemicals used
• Cover crop type (if any)	• Type of wastewater treatment
• Pesticide types and application methods	• Treated effluent reuse (if any)
• Proximity to surface waters	• Other on-site amenities (tourist facilities, restaurants, accommodations, etc.)

assimilate pollutant loadings released during production. As such, when considering the WF of a wine-making process, it could be beneficial to consider the abstracted volumes (green + blue) separately from the assimilative volumes (grey).

Research into the application of the WF process to assess the environmental impact of wine-making on water resources is ongoing. The data needed for comprehensive benchmarking databases are currently lacking, and assessment approaches specific to the wine-making process are still being developed [2]. Despite this, there are a number of opportunities available to utilize this assessment approach in the near-term. For example, the WF process can be used as a tool to assess and quantify the impact of implementing process and operational changes on local water resources. The water audit, a useful tool for reducing process-related water consumption, can be extended to include an evaluation of reductions not only in volumetric water use but also broader environmental impacts. Modifications to operations, such as improving treatment performance and/or beneficial reuse of winery wastewater effluents, may be justified as means to improve sustainability and associated marketability of a wine, despite the costs associated with implementing these changes.

Finally, the results of a WF assessment can be used as a factor in the evaluation of alternative technologies being considered for implementation. Garcia-Alcaraz et al. [23] used both water and Greenhouse Gas (GHG) footprinting to evaluate the environmental impacts of alternative barrel washing techniques. This study highlights the robustness of the WF process as a comparative analysis technique: while different WF methodologies yielded different absolute water consumption and WF estimates, the ratios of these values between cleaning options were similar regardless of the WF assessment approach used. As such, the WF can be a useful tool to estimate the relative impact of various options as part of comparative assessments, allowing selection of a technology or process modification approach that conserves water and reduces environmental impacts.

6 Case Studies

The WF concept is still fairly novel, with the first definitive guidance document published in 2011 [28]. As such, the application of this process to assess the impact of wine-making on freshwater resources is still in its infancy. Despite this, a number of assessments have been completed over the past decade that aim to quantify the impact on freshwater resources from a holistic perspective to a process-specific focus. Table 5 presents summaries of recent WF studies of wine-making processes. Enhancements to the original WF assessment protocol as proposed by Hoekstra et al. [28] that are specific to wine-making have been successfully developed and applied [27, 39]. In all cases, the system boundary was clearly defined, although the breadth of the factors included (viticultural and vinification stages, direct and indirect WF) varied. Typical functional units were a 750 mL bottle of wine and L of wine produced, although some studies also reported values in terms of WF per glass of wine [10, 21].

Using the consumptive approach, the green WF was found to be the largest contributor to the WF of the viticultural stage [39] and, where a LCA was used, the green WF represented the largest contributor to the overall WF [10, 11, 21]. The water-balance approach, by contrast, resulted in negligible values for the green WF and negative values for blue WF, indicating net groundwater recharge within the vineyards of New Zealand [27]. Direct comparison of the consumptive green and blue WFs to those developed using the water-balance approach is not possible.

The impact of drift, runoff and leaching was typically considered during the estimate of grey WF during the viticultural stage, although sometimes only a subset of these factors was considered [39]. Nitrogen was the critical parameter used to assess the impact of leaching on groundwater sources, while pesticides and nutrients in drift and runoff were typically considered. During vinification, simplifications were generally made in the assessment of the grey WF. Lamastra et al. [39] and Bonamente et al. [11] did not consider the impact of WWW on the grey WF, while other assessments assumed a grey WF of, at most, the volume of WWW discharged [10, 21, 27]. A regional assessment of the impact WWW co-treated at municipal WWTPs in Niagara Region, Canada, suggests that the discharge of WWW treated to levels that exceed regulatory requirements can still exert a significant grey WF of as much as 960 times the volume of treated WWW effluents discharged [34].

7 Conclusions

Combined agricultural practices (viticultural stage) and physical and chemical processes (vinification stage) associated with wine-making exert significant water demands and complex impacts on local freshwater resources. Local climatic conditions affect surface and groundwater demands for irrigation, while vineyard configuration and operation affect the need for fertilizer application, runoff volumes and potential for drift. Water is used in large volumes during vinification for cleaning

Table 5 Reported water footprints of wine-making processes

	Region	Scope	Estimation methodology	Water footprint estimate(s)	References
1	Portugal	Vintage stage Direct	Consumptive	Vinification Stage WF_{blue} = 0.15 to 0.20 L/750 mL bottle WF_{grey} = 9.47 to 12.54 L/750 mL bottle	[1]
2	New Zealand—Marlborough and Gisborne Regions	Viticultural and vinification stages Direct and indirect for both vineyard and winery separately	Water-balance (blue) Drift, runoff, leaching (grey)	Viticultural Stage Direct WF_{blue} = −81.3 to −414.9 L/750 mL bottle WF_{grey} = 40.6 to 187.8 L/750 mL bottle Indirect WF_{blue} = 1.0 to 1.1 L/750 mL bottle Vinification Stage Direct WF_{blue} = 2.7 to 3.9 L/750 mL bottle WF_{grey} = 0.5 to 1.1 L/750 mL bottle Indirect WF_{blue} = 10.7 L/750 mL bottle WF_{grey} = 3.3 L/750 mL bottle	[2]

(continued)

Table 5 (continued)

	Region	Scope	Estimation methodology	Water footprint estimate(s)	References
3	Italy—Sicily	Viticultural stage Direct Six vineyards	Consumptive—WFN and V.I.V.A Drift and runoff (grey)	Viticultural Stage WFN WF_{green} = 695 to 903 L/L of wine WF_{grey} = 0 to 229 L/L of wine V.I.V.A. WF_{green} = 705 to 916 L/L of wine WF_{grey} = 0 to 390 L/L of wine WFN-V.I.V.A WF_{blue} = 2.4 to 42.5 L/L of wine	[3]
4	Italy	Viticultural and vinification stages Life cycle assessment, direct and indirect	Consumptive Drift, runoff, leaching (grey)	Overall WF_{green} = 451 L/750 mL bottle WF_{blue} = 120 L/750 mL bottle WF_{grey} = 7.1 L/750 mL bottle WF_{total} = 578 L/750 mL bottle	[4]
5	Romania	Viticultural and vinification stages Life cycle assessment, direct and indirect	Consumptive	Overall WF_{green} = 1,512 L/L wine WF_{blue} = 55.3 L/L wine WF_{grey} = 277 L/L wine WF_{total} = 1,884 L/L wine	[5]
6	Italy—Umbria	Viticultural and vinification stages Life cycle assessment, direct	Consumptive—V.I.V.A. Drift, runoff, leaching, simplification for WWW (grey)	Overall WF_{green} = 621 L/750 mL bottle WF_{blue} = 3.4 L/750 mL bottle WF_{grey} = 7.4 L/750 mL bottle WF_{total} = 632 L/750 mL bottle	[6]

(continued)

Table 5 (continued)

	Region	Scope	Estimation methodology	Water footprint estimate(s)	References
7	Canada—Niagara	Vinification stage Direct Regional	Treated effluent from municipal WWTPs co-treating WWW (grey)	Vinification Stage $WF_{grey} = 1.47 \times 10^7$ to $1.62 \times 10^8 \, m^3/yr$	[7]

Notes Reference numbers refer to the following: 1. [54]; 2. [27]; 3. [39]; 4. [11]; 5. [21]; 6. [10]; 7. [34]

and sanitizing activities, producing a seasonally variable WWW stream requiring handling, treatment and disposal. Where possible, WWW effluents are reused for irrigation, which reduce the volume of water abstracted from the environment for irrigation, however, this practice can potentially negatively affect soil quality. The WF process has been applied to estimate the impact of wine-making on freshwater resources at both the winery and regional levels; however, data are still limited, and differing methodological approaches are still being developed and refined. The consumptive and water-balance approaches yield significantly different results when assessing the green and blue WFs of the viticultural stage. In addition, there is currently no definitive approach in place to quantify the grey WF, particularly during the vinification stage. The current variability in application of the WF concept to the assessment of wine-making limits the ability to use water footprinting as a means to benchmark performance. Despite this, practical applications of the WF approach currently include identifying key process and technological modifications and upgrades that can be implemented to improve product sustainability and marketability. Future research directions could include utilizing the WF approach to model future climate change impacts on local and regional water resources in wine-producing areas, as well as the development of a standardized WF assessment approach that would allow for the creation of a global benchmarking database.

Acknowledgements We thank Natural Sciences and Engineering Research Council of Canada (NSERC) and Ryerson University Faculty of Engineering and Architectural Science Dean's Research Fund for financial support.

References

1. Agriculture and Agri-Food Canada (nd) Canada's wine industry. https://www.agr.gc.ca/eng/food-products/processed-food-and-beverages/processed-food-and-beverages-sector/canadas-wine-industry/. Accessed 8 Aug 2020
2. Aivazidou E, Tsolakis N (2020) A water footprint review of Italian wine: drivers, barriers, and practices for sustainable stewardship. Water 12:396. https://doi.org/10.3390/w12020369
3. Alañón ME, Díaz-Maroto MC, Pérez-Coello MS (2018) New strategies to improve sensorial quality of white wines by wood contact. Beverages 4:91. https://doi.org/10.3390/beverages4040091
4. Allen RG, Pereira LS, Raes D, Smith M (1998) Crop evapotranspiration—guidelines for computing crop water requirements. FAO irrigation and drainage paper 56, FAO, Rome, Italy. http://www.fao.org/docrep/X0490E/x0490e00.htm. Accessed 8 Aug 2020
5. Australian Government (2017) Food Standards Code—Standard 4.5.1—Wine production requirements (Australia only). https://www.legislation.gov.au/Details/F2017C01001. Accessed 10 July 2020
6. Beverage Industry Environmental Roundtable (2012) Water use benchmarking in the beverage industry—trends and observations 2012. https://provisioncoalition.com/Assets/ProvisionCoalition/Documents/Library%20Content/Water%20Management/BIER%20Water%20Use%20Benchmarking%20Report%202012(2).pdf. Accessed 10 July 2020
7. Biddoccu M, Ferraris S, Opsi F, Cavallo E (2015) Effects of soil management on long-term runoff and soil erosion rates in sloping vineyards. In: Lollino G, Manconi A, Clague J, Shan W,

Chiarle M (eds) Engineering geology for society and territory, vol 1. Springer, Cham. https://doi.org/10.1007/978-3-319-09300-0_30

8. Bolzonella D, Zanette M, Battistoni P, Cecchi F (2007) Treatment of winery wastewater in a conventional municipal activated sludge process: five years of experience. Water Sci Technol 56(2):79–87. https://doi.org/10.2166/wst.2007.475

9. Bolzonella D, Papa M, Da Ros C, Muthukumar LA, Rosso D (2019) Winery wastewater treatment: a critical overview of advanced biological processes. Crit Rev Biotechnol 39(4):489–507. https://doi.org/10.1080/07388551.2019.1573799

10. Bonamente E, Scrucca F, Asdrubali F, Cotana F, Presciutti A (2015) The water footprint of the wine industry: implementation of an assessment methodology and application to a case study. Sustainability 7:12190–12208. https://doi.org/10.3390/su70912190

11. Bonamente E, Scrucca F, Rinaldi S, Merico MC, Asdrubali F, Lamastra L (2016) Environmental impact of an Italian wine bottle: carbon and water footprint assessment. Sci Total Environ 560–561:274–283. https://doi.org/10.1016/j.scitotenv.2016.04.026

12. Boulay A, Bare J, Benini L, Berger M, Lathuilliere MJ, Manzardo A, Margni M, Motoshita M, Nunez M, Pastor AV, Ridoutt B, Oki T, Worbe S, Pfister S (2018). The WULCA consensus characterization model for water scarcity footprints: assessing impacts of water consumption based on available water remaining (AWARE). Int J Life Cycle Assess 23:368–378. https://doi.org/10.1007/s11367-017-1333-8

13. Buelow MC, Steenwerth K, Silva LCR, Parikh SJ (2015) Characterization of winery wastewater for reuse in California. Am J Enol Vitic 66(3):302–310. https://doi.org/10.5344/ajev.2015.14110

14. Bustamante MA, Paredes C, Moral R, Moreno-Caselles J, Perez-Espinosa A, Perez-Murcia MD (2005) Uses of winery and distillery effluents in agriculture: characterisation of nutrient and hazardous components. Water Sci Technol 51(1):145–151. https://doi.org/10.2166/wst.2005.0018

15. Christen EW, Quayle WC, Marcoux MA, Arienzo M, Jayawardane NS (2010) Winery wastewater treatment using the land filter technique. J Environ Manage 91:1665–1673. https://doi.org/10.1016/j.envman.2010.03.006

16. Conradie A, Siggee GO, Cloete TE (2014) Influence of winemaking practices on the characteristics of winery wastewater and water usage of wineries. South Afr J Enol Viticult 35(1):10–19. https://doi.org/10.21548/35-1-981

17. Daane KM, Rosenheim JA, Smith RJ, Coviello RL (2013) Western grape leafhopper. In: Bettiga LJ (eds) Grape pest management, 3rd edn. University of California, Richmond, CA pp 202–219. ISBN-13: 978-1-60107-800-1

18. Dharmadhikari MR (2010) What materials are used in constructing fermentors and how does this affect fermentation and storage of wine? In: Butzke CE (eds) Winemaking problems solved. Woodhead Publishing Series in Food science, technology and nutrition, Number 193, Boca Raton, FL. ISBN: 978-1-84569-475-3

19. Dougherty PH (2012) Introduction to the geographical study of viticulture and wine production. In: Dougherty P (eds) The geography of wine. Springer, Dordrecht. https://doi.org/10.1007/978-94-007-0464-0_1

20. Dolder S, Hillman A, Passinsky V, Wooster K (2012) Strategic analysis of water use and risk in the beverage industry. Master's thesis, University of Michgan, Ann Arbor, MI, USA. https://deepblue.lib.umich.edu/bitstream/handle/2027.42/90925/MP_WaterRiskReport_Final_Delivered.pdf. Accessed 10 July 2020

21. Ene SA, Teodosiu C, Robu B, Volf I (2013) Water footprint assessment in the winemaking industry: a case study for a Romanian medium size production plant. J Clean Prod 43:122–135. https://doi.org/10.1016/j.jclepro.2012.11.051

22. Flores L, Garcia J, Pena R, Garfi M (2019) Constructed wetlands for winery wastewater treatment: a comparative life cycle assessment. Sci Total Environ 659:1567–1576. https://doi.org/10.1016/j.scitotenv.2018.12.348

23. Garcia-Alcaraz JL, Montalvo FF, Camara EM, Saenz-Diez Muro JC, Jimenez-Macias E, Blanco-Fernandez J (2020) Comparative environmental impact analysis of techniques for

cleaning wood wine barrels. Innov Food Sci Emerg Technol 60:102301. https://doi.org/10.1016/j.ifset.2020.102301

24. Gomez-Brandon M, Lores M, Insam H, Dominguez J (2019) Strategies for recycling and valorization of grape marc. Crit Rev Biotechnol 39(4):437–450. https://doi.org/10.1080/07388551.2018.1555514

25. Grainger K, Tattersall H (2005) Wine production: vine to bottle. Blackwell Publishing Ltd., Oxford, UK. ISBN-13: 978-14051-1365-6

26. Gu Y, Dong Y, Wang H, Keller A, Xu J, Chiramba T, Li F (2016) Quantification of the water, energy and carbon footprints of wastewater treatment plants in China considering a water-energy nexus perspective. Ecol Ind 60:402–409. https://doi.org/10.1016/j.ecolind.2015.07.012

27. Herath I, Green S, Singh R, Horne D, van der Zijpp S, Clothier B (2013) Water footprinting of agricultural products: a hydrological assessment for the water footprint of New Zealand's wines. J Clean Prod 41:232–243. https://doi.org/10.1016/j.jclepro.2012.10.024

28. Hoekstra AY, Chapagain AK, Aldaya MM, Mekonnen MM (2011) The water footprint manual—setting the global standard. Earthscan LLC, Washington, DC. ISBN: 978-1-8471-279-8

29. International Organization of Vine and Wine (2020) State of the world vitivinicultural sector in 2019. http://www.oiv.int/js/lib/pdfjs/web/viewer.html?file=/public/medias/7298/oiv-state-of-the-vitivinicultural-sector-in-2019.pdf. Accessed 10 Aug 2020

30. Ioannou LA, Puma GL, Fatta-Kassinos D (2015) Treatment of winery wastewater by physico-chemical, biological and advanced processes: a review. J Hazard Mater 286:343–368. https://doi.org/10.1016/j.jhazmat.2014.12.043

31. ISO (2014) ISO 14046:2014—Environmental management—water footprint—principles, requirements and guidelines. International Organization for Standardization, Geneva, Switzerland

32. ISO (2019) ISO 46001:2019—water efficiency management systems—requirements with guidance for use. International Organization for Standardization, Geneva, Switzerland

33. Jackson RS (2008) Wine science—principles and applications, 3rd edn. Academic Press, Burlington, MA, USA. ISBN 978-0-12-373646-8

34. Johnson MB, Mehrvar M (2019) An assessment of the grey water footprint of winery wastewater in the Niagara Region of Ontario, Canada. J Clean Prod 214:623–632. https://doi.org/10.1016/j.clepro.2018.12.311

35. Johnson MB, Mehrvar M (2020) Winery wastewater management and treatment in Niagara Region, Ontario, Canada: a review and analysis of current regional practices and treatment performance. Can J Chem Eng 98:5–24. https://doi.org/10.1002/cjce.23657

36. Johnson MB, Mehrvar M (2020) Charactersing winery wastewater composition to optimize treatment and reuse. Aust J Grape Wine Res 24(6):410–416. h https://doi.org/10.1111/ajgw.12453

37. Karagiannis D, Metaxas T (2019) Innovation in Wine Tourism Businesses: 'Turning Ashes to Gold'. In: Sigala M, Robinson R (eds) Management and marketing of wine tourism business. Palgrave Macmillan, Cham. https://doi.org/10.1007/978-3-319-75462-8_17

38. Kosseva MR, Joshi VK, Panesar PS (2017) Science and technology of fruit wine production. Academic Press, London, UK. ISBN 978-0-12-800850-8

39. Lamastra L, Suciu NA, Novelli E, Trevisan M (2014) A new approach to assessing the water footprint of wine: an Italian case study. Sci Total Environ 490:748–756. https://doi.org/10.1016/j.scitotenv.2014.05.063

40. Liang X, Rengasamy P, Smernik R, Mosley LM (2021) Does the high potassium content in recycled winery wastewater used for irrigation pose risks to soil structural stability? Agric Water Manag 243:106422. https://doi.org/10.1016/j.agwat.2020.106422

41. Lofrano G, Meric S (2016) A comprehensive approach to winery wastewater treatment: a review of the state-of-the-art. Desalin Water Treat 57:3011–3028. https://doi.org/10.1080/19443994.2014.982196

42. Marchal R, Jeandet P (2009) Use of enological additives for colloid and tartrate salt stabilization in white wines and for improvement of sparkling wine foaming properties. In: Moreno-Arribas

MV, Polo MC (eds) Wine chemistry and biochemistry. Springer, New York, NY. https://doi. org/10.1007/978-0-387-74118-5_7

43. Masi F, Rochereau J, Troesch S, Ruiz I, Soto M (2015) Wineries wastewater treatment by constructed wetlands: a review. Water Sci Technol 71(8):1113–1127. https://doi.org/10.2166/ wst.2015.061

44. McRae JM, Teng B, Bindon K (2019) Factors influencing red wine color from the grape to the glass. In: Encyclopedia of food chemistry, pp 97–106. https://doi.org/10.1016/B978-0-08-100 596-5.21655-7

45. Meininger's Wine Business International (2019) Wine adds more that €14bn to Italy's economy. https://www.wine-business-international.com/wine/news/wine-adds-more-eu14bn-italys-economy, April 14, 2019. Accessed 10 Aug 2020

46. Meteo Blue (2020) Climate—Mendoza. https://www.meteoblue.com/en/weather/historycl imate/climatemodelled/mendoza_argentina_3844421. Accessed 10 July 2020

47. Milani M, Consoli S, Marzo Al, Pino A, Randazzo C, Barbagallo S, Cirelli GL (2020) Treatment of winery wastewater with a multistage constructed wetland system for irrigation reuse. Water 12:1260. https://doi.org/10.3390/w12051260

48. Morera S, Corominas L, Poch M, Aldaya MM, Comas J (2016) Water footprint assessment in wastewater treatment plants. J Clean Prod 112:4741–4748. https://doi.org/10.1016/j.jclepro. 2015.05.102

49. Mosse KPM, Patti AF, Christen EW, Cavagnaro TR (2011) Review: winery wastewater quality and treatment options in Australia. Aust J Grape Wine Res 17:111–122. https://doi.org/10. 1111/j.1755-0238.2011.00132.x

50. Mubako ST (2018) Blue, green, and grey water quantification approaches: a bibliometric and literature review. J Contemp Water Res Educ 165:4–19. https://doi.org/10.1111/j.1936-704X. 2018.03289.x

51. Muhlack RA, Potumarthi R, Jeffrey DW (2018) Sustainable wineries through waste valorisation: a review of grape marc utilization for value-added products. Waste Manag 72:99–118. https://doi.org/10.1016/j.wasman.2017.11.011

52. Mulidzi AR, Clarke CE, Myburgh PA (2019) Response of soil chemical properties to irrigation with winery wastewater on a well-drained sandy soil. South Afr J Enol Viticult 40(2):289–300. https://doi.org/10.21548/40-2-3403

53. Pfister S, Koehler H, Hellweg S (2009) Assessing the environmental impacts of freshwater consumption in LCA. Environ Sci Technol 43:4098–4104. https://doi.org/10.1021.es802423e

54. Saraiva A, Rodrigues G, Mamede H, Silvestre J, Dias I, Feliciano M, Oliveira e Silva P, Oliveira M (2019) The impact of the winery's wastewater treatment system on the winery water footprint. Water Sci Technol 80(10):1823–1831. https://doi.org/10.2166/wst.2019.432

55. Shuster D, Paoletti A, Bernini L (2018) Practical guide to grape growing and vine physiology. Board and Bench Publishing, San Francisco, CA, USA. ISBN 978-1-935879-31-2

56. Storm D (1997) Winery utilities: planning, design and operation. The Chapman & Hall Enology Library, Dordrecht, Netherlands. ISBN 978-94-017-5284-8

57. Sturman J, Ho G, Mathew K (2004) Water auditing and water conservation. IWA Publishing, London, UK. ISBN 1-900222-52-3

58. Thornton J (2013) American wine economics—an exploration of the U.S. Wine Industry. University of California Press, Berkeley and Los Angeles, California. ISBN: 978-0-520-27649-9

59. Varela C, Dry PR, Kutyna DR, Francis IL, Henschke PA, Curtin CD, Chambers PJ (2015) Strategies for reducing alcohol concentration in wine. Aust J Grape Wine Res 26:670–679. https://doi.org/ajgw.12187

60. Wang L, Ding X, Wu X (2014) Letter to the Editor: careful considerations when reporting and evaluating water footprint of products. Ecol Ind 41:131–132. https://doi.org/10.1016/j.ecolind. 2014.01.039

61. White RE (2015) Understanding vineyard soils, 2nd edn. Oxford University Press, New York, NY. ISBN: 798-0-19-934206-8

62. Wine Folly (2016) Wine additives: chaptalization and acidification are misunderstood, November 2, 2016. https://winefolly.com/tips/wine-additives-chaptalization-vs-acidification/. Accessed 10 Aug 2020
63. Wines of Canada (2020) Icewine Harvest. http://www.winesofcanada.com/icewine_harvest.html. Accessed 10 Aug 2020

Water Footprint in Leather Tanning and Steel Production

P. Senthil Kumar, P. Tsopbou Ngueagni, E. Djoufac Woumfo, and Kilaru Harsha Vardhan

Abstract Leather tanning and steel production are considered as important sectors in the world market, but their efficiency requires a huge consumption of water. The import and export trade of these industrial activities have been increased in the past decades to deal with the overcrowding population along with the change of lifestyle pattern, as well as the continued industrialization. According to the recent ISO 14,046, water footprint assessment has risen as a key indicator concept of the total freshwater volume consumed and polluted directly or indirectly across a product's end-to-end supply chain. For industry purposes, the focus is paid on the identification and quantification of virtual water trade, the scarcity, and pollution involved in the production of goods and services. To increase the awareness of water consumption in leather tanning and steel production, this chapter emphasizes the impacts of these activities on water use and finally provides a relevant database assessment.

Keywords Water footprint · Leather tanning · Steel production · Water scarcity · Virtual water

1 Introduction

Leather tanning and steel industries are among the most water-intensive industries in the world. While the former stands to be an effective way to recycle raw hides and skins from animals to produce a wide range of goods and located in Mediterranean countries [1], the latter is characterized by a lower cost, high tensile strength and plays a significant role in construction and automobile. Those industries are therefore very important for the development of the economic activities of countries. The products

P. S. Kumar (✉) · P. T. Ngueagni · K. H. Vardhan
Department of Chemical Engineering, Sri Sivasubramaniya Nadar College of Engineering, Chennai 603110, India
e-mail: senthilkumarp@ssn.edu.in; senthilchem8582@gmail.com

P. T. Ngueagni · E. D. Woumfo
Laboratoire de Chimie Inorganique Appliquée, Faculté Des Sciences, Université de Yaoundé I, Yaounde, Cameroon

© The Author(s), under exclusive license to Springer
Nature Singapore Pte Ltd. 2021
S. S. Muthu (ed.), *Water Footprint*, Environmental Footprints and Eco-design of Products and Processes, https://doi.org/10.1007/978-981-33-4377-1_5

derived from the leather industry come from renewable and available resources of the slaughterhouse. However, the conversion of raw material into a finished product involves the usage of many chemicals and water in all the production chains to satisfy a global demand especially in the faster-developing countries regions of Asia and Latin America [2]. Additionally, the chemicals used are generally toxic and hazardous to human and water bodies. For a few decades, the global consumption of steel is symbolized by an upward trend and is the basis of several industries in the world. This industry enhances the competitiveness of national manufactures and the growth of the national economy. China has been pointed out as the world's largest producer of iron since 1996, thus influences the international development of the steel sector [3]. Notwithstanding, China suffers from water scarcity and 32% of water encounters different pollution problems. To deal with this environmental problem generated from leather and steel industries, water footprint assessment (WFA) derived from life cycle assessment has been developed recently. According to the International Organization Standard (ISO), the influence related to water quality should be considered in water footprint analysis [4]. Likewise, WFA takes into account direct and indirect processes along with the concept of virtual water. Water is undoubtedly one of the most important natural resource used in industries processing in the world. In order to satisfy a huge demand for goods and services, and also export some commodities across the world, many countries through their industries consume a significant quantity of virtual water. In the leather industry, products are characterized by their toughness and durability which is an essential clue for attracting consumers. However, from hide to leather, the tanning process generally requires an extensive amount of water, which is associated with some pollutants including heavy metals, dyes, sulfur, and surfactants. Then, the expansion of tanneries which represent a livelihood for many people in developing countries may be an environmental burden in terms of pollution and water scarcity.

This chapter is therefore an attempt to show the systematic assessment of water discharge from leather and steel industries, present the parameters to be taken into account, and give recommendations for better sustainability of water resources.

2 Leather Industry: Concepts and Processing

The leather-based industry is an important segment of the economy worldwide, and a very old manufacturing sector producing a broad range of products including leather garments, leather footwear, leather bags, etc. In the leather industry, the crude material is generally derived from the waste product of the food industry, exclusively from meat processing in slaughterhouses [5]. The resource is then found to easily available and renewable. According to the recent study carried out by the Food and Agriculture Organization of the United Nations, over 6.6 million of bovine hides and skins were produced in 2014 [6]. Leather can then be considered as one of the most widely traded commodities in the world (it is ranked among the 25 traded in the world), and its industry holds an outstanding place in many countries particularly in Mediterranean countries. For instance, due to the increase of population and the uttermost urbanization in countries, the current trade presently outpacing US$ 80 billion in a year and will increase in a few years. Slaughterhouses and waste from the meat industry are the primary raw material used during the leather processing industry. In common tanneries, leather processing involves many steps, with the use of various chemical and mechanical processes to transform hides and skins into higher valuable material through the tanning process.

The tanning process can be defined as a conversion of raw hides and skin of animals into a stable, flexible, useful, and desirable material suitable for a wide range of applications. Although recent studies have revealed different phases of this process [7], most of the authors conversely suggest three broad processes classified into the Preparatory or beam house stage, tanning, and post-tanning operations.

3 Beam House Stage Activities

Many operations are involved in this first to make the hide/skin ready for tanning. Although many pretreatment options exist, the most common may include the following:

Preservation

In this operation, untanned hides are directly prepared for tanning or preserved by curring (to prevent putrefaction of the protein substance from bacterial growth), salting, and freezing. This treatment is therefore useful to protect animal skin from any decomposition.

Soaking

Due to the high content of salt used in the previous step, the soaking consists in to remove the absorbed salt, bring the hides and skins back to their initial condition, rehydrate and wash the skin, and finally remove unwanted materials such as blood, manure, urine, dung, dirt, and interfibrillar proteins for subsequent processing. During this step, major raw material need 6,000–9,000 L water/t hide to be clean up.

Green Fleshing

This step stands for mechanical removal of fats and flesh from the inside of skins or hides. Based on the environmental concept, green fleshing is operated when the tissue is not contaminated with chemicals, then the number of chemicals required for the next process will be reduced.

Liming

The liming is performed in two stages. The first called dehairing operation concerned the mechanical removal of wool and flesh. The second step designated as reliming consists of the diffusion of chemicals in the hide matrix of consecutive operations for further elimination of protein and opening-up process. The hides are finally fleshed to remove subcutaneous material and weighed. Sodium sulfide, lime, and/or an enzyme is used during the dehairing stage, whereas only 3–3.5% is used during the reliming process. The final product obtained after the liming operation consumes 4,000–6,000 L water/t hide.

Deliming-Cum-Bating

In order to neutralize and remove the lime from pelt, ammonium sulfate or ammonium chloride is used. Phenolphthalein is used indicator to confirm the elimination of lime in water by the absence of pink color. This step is following by the bating stage which is the treatment of delimed hides with enzymatic precaution for the removal of short hair, scud and enhances grain characteristics. During this process, 4,500–5,000 L water/t is used.

Pickling

Pickling is defined as a process of conditioning delimed and bated pelts for chrome tanning. A pH in the range of 2.8–3.0 is required to reach the equilibrium by using sulphuric acid. This helps to bring the pH of collagen to a very low level to ease the penetration of mineral tanning agent into the substance. A range of volume of 800–1,000 L water/t is used during this process.

4 Tanning Process

Tanning is a chemical process of converting raw hides or skins into leather through stabilizing agents against heat and enzymatic attack [8]. Also, this process tends the raw material to change their chemical, physical, and mechanic properties. Different agents such as aldehydes, synthetic tannins, mineral tanning salts, vegetable tannins materials are used to stabilize hides or skins. The most common tanning methods are chrome, vegetable, or a combination of both. The choice of a method and any agent depend on raw material, cost of materials, and the quality of the final product. In the chromium process, chromium(III) sulfate is ($[Cr(H2O)_6]_2(SO_4)_3$) has always been used as the most adequate tanning agent. The increase of pH in a particular range is being possible by using sodium bicarbonate leading to a cross-linkage between the collagen and the chromium. In vegetable tanning, tannins naturally occur in the bark or leaves of many plants are used as a class of polyphenol chemicals. Collagen is then coated by tannins, render them less-water soluble and more resistant to bacterial attack, finally, the hides are more flexible. This second step consumes around 1,500–2,000 L water/t pelt. Nevertheless, recent studies have stated the chrome tanning as the most used, but chromium tends to be noxious for human health when its uptake value is beyond 13.4 μg 1,000 kcal^{-1} [9], and also for various types of cancer [10].

5   Post-Tanning Operations

This last stage is also considered as the finishing process which consists of rechrome tanning, neutralization, retanning, dyeing, and fatliquoring.

Rechrome Tanning

In this step, a basic chromium sulfate is used to treat and level the tanned hides. A volume per chrome-tanned leather produce is between 2,000 and 3,000 L.

Neutralization

The neutralization consists of the reduction of the cationic surface charge of chrome-tanned leather so that the uptake of dyestuffs, fat liquors, and anionic retanning agents will be monitored. A better and uniform distribution of processing agents during successive steps is then achieved. A total volume of 5,000–6,000 L water/t tanned leather is required to accomplish the neutralization process.

Retanning

Due to the lessened molecular weight of chrome complexes, the tanned leather produced are generally not saturated, then retaining is always required. Different retannage and retanning materials are available, and are used together or by combination according to the final of the desired product. Acrylic syntan, vegetable tannins (quebracho, wattle, phenol–formaldehyde condensates), and resins syntan based on melanine are the various retanning materials generally used.

Dyeing

Leather dyeing is a transition process between tanning and finishing. This process is in relation to retanning and fatliquoring, as well as these materials are anionic in nature, and one can act as a regulator to the other. The final physical properties required in the leather such as light fastness, rub fastness, resistance to solvents, perspiration resistance, etc., determine the choice of dyestuffs.

Fat Liquoring

In this step, leather is lubricated and softens. The nature of substrates, the stage of the application, the nature of the fat liquor, and the type of finished leather are the parameters to be considered for the affinity of fat liquor to tanned leather and its efficiency in lubricating fibers. Generally, retanning, dyeing, and fat liquoring are implemented together, with a water requirement of 2,500–3,500 L/t tanned leather (Fig. 1).

The leather industry is therefore a sector of both public and private economies on which many countries depend. Also, in agreement with the Food and Agriculture Organization (FAO) of the United Nations, the interest in leather products, in general, is expected to continue in the world especially in developing countries as presented by recent findings throughout different regions in the world [6]:

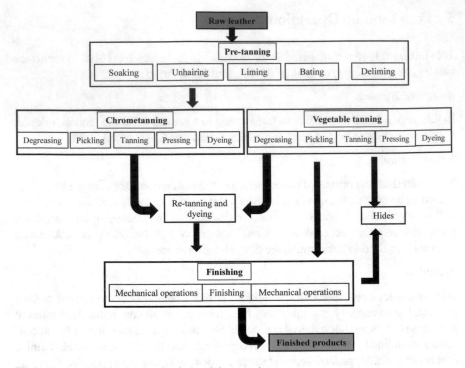

Fig.1 Flowsheet of leather tanning industrial processing

- Far East: 285 10^3 t/year of bovine leather production with 225 t/year of sheep and goat production
- Latin American and Caribean: 110 10^3 t/year of bovine leather production with 16 t/year of sheep and goat production
- Africa: 5 10^3 t/year of bovine leather production with 49 t/year of sheep and goat production
- North America: 21 10^3 t/year of bovine leather production with 6 t/year of sheep and goat production
- Area Former USSR: 38 10^3 t/year of bovine leather production with 22 t/year of sheep and goat production
- Europe: 71 10^3 t/year of bovine leather production with 73 t/year of sheep and goat production

Likewise, during the process of leather tanning, some by-products such as wool are useful for textiles manufacturing; Leatherboard may be produced from tanyard splits and trimmings, whereas collagen may contrarily by manufactured from beamhouse splits and trimmings.

Nevertheless, an increase in leather production leads to the overexploitation of water resources as well as the pollution of the environment due to the use of chemicals. Then, the measurement of water footprint may be a crucial path to quantify the direct and indirect water usage by consumption or manufacture of leather and reduce the consumption of water.

6 Pollution Derived from Tanning Process of Leather

It is well known the economic impact of leather industry on the development of a locality, and extended to a country. Nonetheless, the different activities that generated solid waste and wastewater effluent consisted of biocides, synthetic tannins, detergents, and resins [6, 11]. As previously presented in the different tanning process, the leather industry uses a huge amount of water and the processing generated pollutants to the environment. During the different stages of leather tanning processing including soaking, unhairing/liming, deliming and bating, chrome tanning, post-tanning, and finishing, the polluting load from produced wastewater consist of sulfates, chlorides, Chromium, sulfides, suspended solids, etc., at various ratios. It has been noticed from previous studies the high impact (nearly 90%) of pretanning and tanning on the overall pollution from tannery [5]. As far as concerned the pretanning process, the enhancement of chemical species such as chlorides, sulfates, total dissolved solids (TDS), chemical oxygen (COD) in tannery wastewater are mainly due to the fluctuation of pH. The parameters including biochemical oxygen demand (BOD), COD, and suspended solid result in the dehairing process by using sulfide and lime. Moreover, the negative effects of sodium sulfide due to its strong alkaline properties as found to be harmful both to the environment and plants. The use of chromium during chrome tanning especially generates the shortcomings of the ecological system. Finally, the waterbodies may be altered by the exceedingly discharge chemicals from polluted sediments.

7 Wastewater Treatment Processes in the Leather Tanning Industry

In the leather industry, the wastewater treatment systems consist of four main steps including pretreatment or preliminary treatment, primary, secondary, and tertiary treatment.

Pretreatment.

Mechanical screens, distribution wells, grit removal apparatus, and equalization tanks are used in this step in other to separate suspended solids from raw waste and to remove solid material (hair, coarse, sand, grit, and grease).

Primary treatment.

Two different processes are considered in this step including the following:

- Physical separation of fats
- Easy sedimentation of suspended solids in primary settling tanks by using coagulants and flocculants

During this step, chrome and sulfides are precipitated from tanyard chrome and beam house; In order to deal with the hydrogen sulfide produced, oxidation using a metal catalyst is possible and must be operated separately to avoid.

Secondary treatment.

This step is an anaerobic biological treatment to reduce the biological oxygen demand (BOD). The treatment is suitable for the reduction of sludge at lower costs and is extensively used in industries. However, the toxicity and corrosive properties of anaerobic are the drawbacks faced by many tanneries that have not been adopted.

Tertiary treatment.

This step consists in using sand filter beds and clarifiers to separate the liquid effluent from the solids and dewatering the sludge. Even though the drying process of sludge is more recommended, the liquid effluent produced is unfortunately released to the environment.

8 Toxicity of Some Chemicals Used During Leather Processing

During the manufacturing process of the leather, an extensive range of chemicals are used. Among the bunch of chemicals, one can cite.

- Phtalates such as benzyl butyl phthalate (BBP), di-ethyl hexyl phthalate (DEHP), and dibutyl phthalate (DBP) are employed as plasticizers in microporous artificial
- Biocides: Used during the finishing process of microbiological water-based chemicals. The combination of biocides such as chlorisothiazolinone (CIT) and methyl-isothiazoline (MIT) can be used as irritant [13]
- Formaldehyde: It acts as a cross-linker and helps for the manufacture of diverse type of leathers
- Inorganic pigments: Their properties including brilliant color and fastness is required especially for chromate and lead, although considered as toxic species [12]
- Organotin compounds: Used in the finishing process like dibutyl tin
- M-Methyl pyrrolidone acts as a wetting agent, swelling material, and plasticizer to improve the performance of leather

It is noted that a bunch number of chemicals used during the tanning processing, which after discharged into the environment will be harmful to human health and plants.

9 Water Footprint Assessment in Leather Tanning Processes

The development of root concept and different water footprints have enhanced the environmental thoughts of water resources for the better promotion of its sustainability. Water footprint has been considered as a consumptive product that has early been studied but also degradative freshwater use. For instance, along with the concepts of blue and green water footprints previously well described, the notion of greywater footprint has been recently introduced. Green water footprint measures the volume needed to assimilate pollutants entering freshwater bodies [13]. In this conceptual framework, water footprint doesn't only consider the consumption of freshwater, but also the pollution and scarcity, factors which can generate hazardous products for human and the environment. Broadly spoken, in order to deal with the traditional management of wastewater centered on the deepening role of government, water footprint assessment is an opportunity to show the responsibilities of all the production chain actors to preserve the water resources. Based on this thought, water footprint assessment can be easily applied to any industry like leather. In fact, in Asia and Africa where there is a huge livestock potential, the leather industry remains one of the most economic sectors which will be improved in a few years. Then the need to evaluate the water footprint from all the components of leather is necessary.

The leather industry is one such industry that produces products that have a huge value of virtual water. The water footprint assessment of the leather industry depends on its uptake of raw materials and the quantification of its products. Generally, leather consumes 30 m^3 of water per ton of leather in the process. However, 6% is consumed and 94% is released into the environment as wastewater. In order to deal with the problem of chromium content in wastewater effluent, the combination of tanning using vegetable tannins and aluminum sulfate is appropriate. In fact, aluminum is abundant on earth, and vegetable tannin is a renewable resource. Likewise, the use of both materials produces a final product with high quality nearly to similar to chrome tanning.

It is well known the leather tanning is one of the most-intensive industries worldwide. For instance, it uses a huge amount of water for animal raising, but also from tanning finishing processes. The reduce, reuse, and recycle stands to be a significant path to deal with water pollution and scarcity. The water footprint of the leather industry should be estimated not only during the tanning but also from animal raising. It has been noticed that 1,890,000 L of freshwater is needed to raise one cattle, with approximately 55% can be attributed to the leather leading to a value of 103,950 L for a single hide.

Another important factor to be considered in the leather industry is that green water prevails the overall water footprint, representing around 83% of the total water footprint during the treatment of raw hides. The quantification of the water footprint of the leather industry involves the water footprints of all the raw materials. During this process, the use of a wide range of chemicals with various sources tends the process difficult by quantifying the water footprint of non-raw materials. Conversely, during the tanning process, water use is usually released back into surface water or percolates the groundwater. There is therefore a loss in terms of water resources for other purposes.

10   Waterless Chrome Tanning Technology: A Breakthrough to Reduce Water Footprint from the Leather Industry

Wastewater generated from leather tanneries is undeniably one of the most polluted water bodies used chromium. The chromium content in those water is at a high level, and unfortunately expensive to remove. New technology has been developed a few years ago in India to deal with this problem called waterless chrome tanning technology (WCTT). Almost 2 billion square feet of leather is produced in India, and 70,000 tons of basic chromium are used annually. Practically, the absorption of chromium tanning agent unabsorbed gets released in addition to wastewater. Then, to promote the environmental sustainability of leather processing and reduce the amount of water used, the waterless chrome tanning technology is a good approach that has attracted the interest of several countries in the world including Brazil, Vietnam, New Zealand, South Africa, and Ethiopia. This technology displays many benefits and outlines including.

- Complete elimination of water input for chrome tanning
- Simply handling process without the use of infrastructures nor chemical
- Reduction in water usage
- Suitable for all type of products varieties from hides and skins
- Reduce usage of chromium by 15–20% deriving from material saving
- Shorten the total dissolved solids in wastewater from this process by 20%
- Total elimination of pickling and basification processes by enhancing the productivity

This new technology is therefore a good path to reduce water pollution and scarcity, will also positively impact the water footprint of the leather industry and may be translated in worldwide.

11   Steel Production in the World

Though using nowadays to build our new modern world, steel production has started thousands of years back for the development of metallurgy in sub-Saharan African countries and the Roman Empire as well. Steel has known deeply changes with new production techniques especially in India (around BC) and in China (first century AD) through the international steel market. Steelmaking has also known an exponential development after the industrial revolution in Europe and North America, following by a significant demand to trade and transport such as shipbuilding and railways. Nowadays, steel is considered as one of the essential world's materials. Its industry employs more than 2 million people worldwide industries. Transport, automotive and construction power and machine are the most industries relying on steel production because of its low cost and tensile strength. World crude steel production has been increasing in the world by the development of new products, improvement of the manufacturing process, and future breakthrough technologies.

Despite the awareness given by the Organization of Economic Co-operation and Development (OECD) concerning the raise of 55% of global water abstraction between 2000 and 2050, it is noted that many countries especially in arid geographic zone strategically depend on steel production which is one of their most income revenue [16]. The steel industry is significant in energy consumption, as well as freshwater. Steel products are generally extracted all over the world, conveyed and spread again, then responsible for the consumption of global resources of freshwater. Therefore, it is necessary to study the steel production chain in order to evaluate the volume of water used during the processing and assess the water footprint deal with the water consumption problem. The steel industry has an important impact on the economy of many countries in the world and is considered as a major pillar. Global steel production has raised from 2008 to 2010 and China was the first producer following by Japan, America, Russia, and India [14].

In fact, according to the World Steel Association analysis, global steel production raised from 1.348 tons in 2007 until 1.669 tons in 2014. By 2017, the production has reached 1.691 billion tons of crude steel representing 5.3% of the production growth in relation to 2016. Equivalent numbers were found in Europe representing 313.2 million tons in 2017, which counts for 3.8% growth from the 2016 steel production level. In European Member states, the 4.1% yearly growth accounted for 168.7 million tons in 2017, after 210.6 million in 2007 and 162 million in 2016 [18]. These recent findings showed the continuous demand for steel for our society which might be due to the increase of innovativeness, quality, and customer orientation. As far as steel production is related to the consumption of freshwater, it is important to point out that this industry seriously damages local, regional, and global water resources. The wastewater coming from the steel industry generally contains harmful pollutants including dissolved metals like cadmium, arsenic, petroleum-derived products, volatile phenol, etc.

12 Production Chain of Steel

On the basis of underground and surface mineral deposits, the production of steel takes place on different routes and depending on countries. For example, in China, an electric arc furnace (EAF) and the integrated steelmaking (BF-BOF) are the main steel production routes. This last route accounts for 95% of the national steel products in the country. However, two main routes are usually used to produce steel:

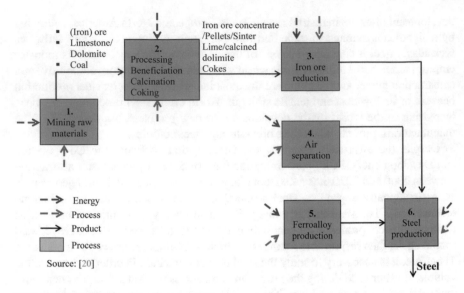

Fig.2 Steel production chain. Source: [19]

- Blast furnace (BF);
- Basic oxygen furnace (BOF).

The former is a furnace where oxygen is removed from iron ore by binding it to carbon, whereas the latter is a furnace where the carbon content in the iron is lowered by blowing pure oxygen onto the metal. The steel production chain consists of six steps as represented in Fig. 2.

As it is presented in Fig. 2, the steel production chain requires energy and water for the overall functioning process for expected products. Each step is particular in terms of material and some processes.

First step: Some raw materials like limestone ($CaCO_3$), dolomite ($Ca(Mg(CO_3)_2)$), coal, and other ores for alloyed steel, including chromite and laterite are mined.

Second step: An improvement of raw materials is taking place during this step by the different processes.

- Beneficiation: Here, there is an increase in ore concentration and fine ore particles are bound to form sinter or pellets. During this process, water is used for dust emission control, sorting material, cleaning, gas treatment, and cooling.
- Calcination: Dolomite and limestone are used here to produce lime (CaO) and calcined dolomite (CaO, MgO). Water is sometimes used to wash limestone.
- Coking: Coal properties are enhancing during this process to produce cokes which contain higher carbon purity. The final obtained cokes are subjected to a wet quenching by water.

Third step: New materials such as pig iron and reduced iron oxide are formed by putting iron coke, cokes, and limestone in the blast furnace. Water is used for blast furnace gas treatment, slag granulation, and cooling.

Fourth step: Air separation, oxygen for steel production in the basic oxygen furnace is produced by separating oxygen from the air. The metal–carbon is lowered by blowing pure oxygen over the hot metal. Water is required for cooling and electricity provides the energy required for separation.

Fifth step: It is noted that stainless steel production mostly required requires ferrochrome and ferronickel. In this step, iron and other metals are added to the basic oxygen furnace to produce alloyed steel. Water is employed for gas treatment, slag granulation, and cooling.

Sixth step: The materials resulted from step 3 (pig iron from the iron ore reduction process), roughly contains 4% carbon, are transferred to the basic oxygen furnace carbon-reducing by blowing pure oxygen onto the hot metal. Also, water is required in this step for gas treatment, vacuum generation, cooling, and washing [19].

Based on the above steel production chain, it is well noted that energy and water sustainability are undeniably intertwined, and then required a huge quantity of water in all the six compartments.

13 System Boundary in Steel Industry

The industrial sector is the second major water consumer after agriculture in the world. Then, in order to reduce water consumption and virtual water, the designing of a suitable system boundary is very important. A system boundary can be defined as a scheme that takes into account the different inputs and materials involved during the processing to obtain the final products by assessing the pollution generated by wastewater. The illustration of a system boundary depends on each study and the components to be considered. In a recent study concerning the life cycle of water used and wastewater discharge of steel production in China, the system boundary was regrouped into the steel production layer, steel enterprise, and social environment [15]. The environment in this case was encompassed in the boundary because the water treatment footprint concept also takes into account the pollution phenomena. These three levels of steel boundary include sintering, coking, iron making, pelletizing, hot rolling, and cold rolling processes. It was noted that the social environment layer includes purchased electricity and upstream intermediate products (oil, coal, coking coal, purchased sinter), whereas the steel enterprise boundary comprises auxiliary processes, electricity, and wastewater treatment plants. Contrarily, a system boundary in a steel industry consists of iron steelmaking, continuous cast, steel rolling and other processes, water consumption of staff, mechanic energy costs, chemicals use, and transportation was used in another study [14]. Although the life cycle can be extended to a country, spatial boundary analyses in literature are focused on particular enterprises, and enhance a methodology to develop a common and appropriate water footprint assessment.

14   Life Cycle Assessment (LCA) Method

Life cycle assessment (LCA) is a methodological framework for estimating the environmental impacts attributable to the life cycle of a product, including climate change, stratospheric ozone depletion, tropospheric ozone (smog) creation, eutrophication, acidification, toxicological stress on human health, and ecosystems, the depletion of resources, water use, land use, and noise [21]. In the steel industry, the LCA method is suitable to evaluate the total environmental impacts of products from cradle to grave. Much advancement has been developed years ago to provide a framework of LCA based on international and draft standards of ISO 14,000 series. Those standards included the following:

- International standard ISO 14,040 (1997) on principles and framework;
- International standard ISO 14,041 (1998) on goal and scope definition and inventory analysis;
- International standard ISO 14,042 (2000) on life cycle impact assessment;
- International standard ISO 14,043 (2000) on life cycle interpretation.

As an efficient tool, LCA considers a wide range of environmental issues such as the discharge of pollutants and climate change to potentially quantify the impacts on products at different stages of their lifecycle. The research is concerning the role of LCA in water in 2009 [19], following a great interest in water use in LCA databases. Water is then considered in life cycle assessment as an input from nature in which the utilization can develop an environmental impact related to known production activity. LCA therefore mainly focuses on the extraction and manufacturing processes for a better understanding of environmental impacts. In the steel industry, LCA mat help scrutinizing the elements of the environmental factor correlated among corporate activities; Likewise, LCA can further help to come upon the root causes of different environmental issues associated with products, overcome their management drawbacks, and accomplish the entire process control.

15   Impacts of Water Footprint on Steel Industry

The ISO 14,026 related to the principles of requirements and guidelines on water footprint was promoted to deal with the impacts of water scarcity. In order to well understand the role of LCA in a water footprint, the development of a suitable methodology outline is very important. According to the international standards, the different steps to be considered include the following:

- The definition of goal and scope: Describe the functional unit and system boundaries;
- Inventory analysis: Describe the life cycle steps, gather all the inputs and outputs elements, and evaluate the components of each element;

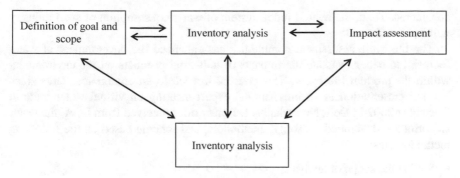

Fig.3 Life cycle assessment steps

- Impact assessment: Determine the environmental impacts that are type and intensity of pollution;
- Interpretation and improvement: Compare the results of the analysis and give suggestions for environmental impacts.

The different LCA steps are therefore intertwined and may be represented as follows (Fig. 3):

The steps of LCA and its entire methodology therefore serve as a key to enhance and improve the understanding of environmental impacts of water footprint which can be useful for steel industry.

16 Water Footprint Based on LCA

16.1 Water Footprint Notions, Methodology, and Formulation

Four notions are generally considered on water footprint assessment. The first notion concerns the assumption that freshwater is a global resource used by people anywhere. The second notion is related to the limited regeneration rate of freshwater. Indeed, when used for a precise activity, freshwater can be reoriented elsewhere. The third notion is based on the entire consideration of production chain. The last notion advices to recognize both water consumption and water pollution. Steel industries sometimes require freshwater consumption which is freshwater used to produce 1 ton of steel of a factory (FWC) summed to water system of steel industry, without counting water for cooling; Also water consumption (WC) per ton of steel which is water used to produce 1 ton of steel by ignoring reclaimed and recycled water). Freshwater is assigned to fresh tap water, surface, or groundwater. However, the simple calculation of FWC and WC is not in agreement with the practical condition because water used to produce inputs in the process referring as virtual water is not considered. Then, the overall index system of water consumption of steel industry will be re-evaluated.

The life cycle assessment community has enhanced the development of water footprint to better evaluate the impacts of industrial products on the environment within the product life cycle. This concept has led to an emergence status since his first presentation in an international expert meeting on virtual water trade in December 2002 in Deft, Netherlands. Despite critics received from LCA, the water footprint has developed a strongly methodological scheme based on the following methodologies:

- Setting the scope of analysis;
- Accounting;
- Sustainability assessment;
- Response formulation.

It is therefore well known although life cycle assessment and water footprint have different roots, they can be put together to efficiently deal with water footprint in steel industry.

The summation approach and the stepwise accumulative approach are the most common methods used to evaluate the water footprint. Though water footprint comprises gray water footprint, blue water footprint, and green water footprint, their simple summation although bringing useful information for only one product, is not environmentally suitable for manufacturers. Moreover, industries generally don't possess their rainwater harvesting system, the green water is therefore not considered. However, the stepwise accumulative water refers to a general calculation of water footprint depending on the water footprint of the final steps in the production of final and required products and also on the water footprint calculation in the processing steps [19], then suitable for complicated products chain like steel industry. The different elements of a steel production chain (smelting refining, continuous casting, rolling along with other processes) are interrelated, so their water used should be individually taken into account.

The water consumption footprint can be calculated by the following formula:

$WCF = DWF + VWF.$

where WCF is the water consumption footprint, DWF is the direct water footprint, and VWF is the virtual water footprint. The direct water footprint can also be obtained by.

$DWF = WF_{obtained} - WF_{D\text{-}discharge} - WF_{loss}.$

where $WF_{obtained}$ is the amount of water obtained, $WF_{D\text{-}discharge}$ is the amount of direct water discharge, and WF_{loss} is the loss caused by evaporation, infiltration, and by-products.

In steel industry, water footprints consist of natural gas mining and processing, electricity generation, coal mining, iron ore mining, limestone mining, transportation, and employees' meal which water footprint values are presented as follows [17]:

- Natural gas mining and processing: 9.251×10^{-3} (m^3/m^3)
- Electricity generation: 1.8 m^3/MWh

- Coal mining: 0.36 (m^3/ton)
- Coke processing: 1.03 (m^3/ton)
- Iron ore mining: 0.425 (m^3/ton)
- Limestone mining: 1.09 (m^3/ton)
- Transportation: 0.22 × 10^{-3} (m^3/km)
- Employees' meal: 4.757 (m^3/meal)

However, compared to the easier quantification of the virtual water footprint of domestic use, the virtual water footprint of a steel industry is complex to evaluate. Indeed, the information concerning the water for production inputs, internal electric power consumption, transportation, chemicals, and domestic water consumption is generally compulsory, but not unavailable.

It is also very important to point out that there are three major water risk assessments based on water footprint. Those risks include.

- Regulatory risk;
- Physical risk;
- Reputation risk.

Industries generally faced physical risk which is based on water resources when subjected to water shortage or pollution. The water footprint is an appropriate tool in water risk assessment of water and combines three considerable elements including water risk management, water risk assessment, and water footprint calculation.

17 Water Footprint Measurement and Recommendations

The water footprint of steel industry is very important compared to other related industries. Their sewage discharge is generally harmful to the environment. The components of steel industry sewage include lead, cadmium, arsenic, zinc, chloride, chromium, nitrate, volatile phenol, petroleum, etc., which are very difficult to treat and may negatively affect the local water quality. In steel industries, the WFA is generally the simple summation of green water, blue water footprint, and gray water footprint. However, the gray water affects the consideration of environmental water footprint assessment. Also, the virtual water footprint is closely linked to risk analysis which can help manufacturers to consider gray and blue water footprints. The recent study water footprint has shown that the total water pollution was 27 times of blue water footprint [14], despite the wastewater treatment system plant of industries. So, the reduction of the gray water footprint must be prioritized. In the same study, authors have pointed out the limiting of water resources as the main drawback of steel industry. Additionally, water risk assessment has helped to categorize risk of pollution, risk of the supply chain, risk of energy, and risk of water consumption to deal with and enhance water management and sustainable development by steel industry. The treatment of sewage is, therefore, a key to efficiently reduce the gray water footprint of steel industry. Also, environmental-friendly chemicals such as green agents (corrosion scale inhibitors) are suggested. The improvement of their productivity processes like internal recycling, treatment of cooling water, and the use of sophisticated equipment is also recommended. Though the management of water system must be enhanced, rainwater harvesting is an opportunity of green water for industries. In order to avoid water pollution phenomena from wastewater discharge, use gray water in steel industry, advanced wastewater treatment plants must be needed to encourage their reuse. Steel producers need to reduce direct water consumption by enhancing the efficiency and upgrading the equipment. The processes including cooling, agglomeration, and pelletizing which use a high amount of water will be reduced. The use of renewable energy sources such as wind power is also highly recommended and a priority should also be given to the system plant location [23]. However, some uncertainties about water footprint calculation remain unanswered because of the lacking of reliable data. For instance, steel industry component consumes a large amount of power electricity generated by virtual water which can be reduced by knowing information on the water footprint of local or regional electricity generators. As far as concerned gray water is used to dilute pollutants to be discharged in the natural water system, industries must gather information to better evaluate their volume. Finally, the reliable calculation of the water footprint of steel industry must be done separately to better evaluate the volume and the impacts of each component.

18 Conclusion

The water footprint is a key factor to better evaluate the direct and indirect water generated from leather tanning and steel industries. The definition of a system boundary is important during different processes. The impacts of pollution on humans and the environment have been presented in this chapter. As far as concerned the leather industry the use of waterless chrome tanning technology has been displayed as a breakthrough to deal with water pollution and reduce the volume of water during the processing of raw hides and skins. The production of steel has been increasing all over the world. Different notions and formulation to evaluate the water footprint of this industry have been presented. Independently on the steel production routes, water in the steel industry needs cooling water to produce electricity for power plants. Additionally, operations including washing, furnace gas treatment, vacuum generation, and slag generation also require a huge quantity of water during the processing of iron mining. The volume of water used in steel production is very large and deeply affects the water footprint. Different recommendations have been suggested including the usage of renewable energy like power wind and upgrading equipment during the processing to reduce the water footprint of steel industry.

References

1. Lafrano G, Meric S, Zengin GE, Orhon D (2013) Chemical and biological treatment technologiesfor leather tannery chemicals and wastewaters: a review. Sci Total Environ, 461–462, 265–281
2. UN-FAO (2013) World statistical compendium for raw hides and skins, leather and leather footwear 1993–2012, Trande and, 193
3. UNESCO (2015) The united nations world water development report. The United Nations Educational, Scientific and Cultural Organization
4. International Standard, ISO 14046 (2014) Environmental management-water footprint-principles, requirements and guidelines, International Organization for Standardization, Geneva, Switzerland
5. Dixit S, Yadav A, Dwivedi PD, Das M (2015) Toxic hazards of leather industry and technologies to combat threat: a review. J Clean Prod 87:39–49
6. FAO (Food and Agriculture Organization of the United Nations) (2015) World statistical compendium for raw hides and skins, leather and leather footwear 1998–2014. I4651E/1/05.15. Food and Agriculture Organization of the United Nations (FAO), Rome
7. Giaccherini F, Munz G, Dockhorn T, Lubello C, Rosso D (2017) Carbon and energy footprint analysis of tannery wastewater treatment: a global overviem. Water Resour Ind 17:43–52
8. Beghetoto V, Zancanaro A, Matteoli U, Pozza G (2013) The leather industry: a chemistry insight part I: an overview of the industrial process
9. Jin J, Mulesa L, Rouillet MM (2017) Trace elements in parental nutrition; Considerations for the prescribing clinician. Nutrients 9:440
10. Yan J, Huang H, Liu Z, Shen J, Ni J, Han J, Jin L (2020) Hedgehog signalling pathway regulates hexavalent chromium induced liver fibrosis by activation of hepatic stallate cells. Toxixol Lett 320:1–8
11. FAO (Food and Agriculture Organization of the United Nations) (2015) World statistical compendium for raw hides and skins, leather and leather footwear 1998–2014. I4651E/1/05.15. Food and Agriculture Organization of the United Nations (FAO), Rome
12. Jerry S (2011) Environmental production and public health issues, 88. Leather, fur and footwear. In: McCann M, Stellman JM (eds) Encyclopaedia of occupational health and safety. International Labor Organization, Geneva
13. Shakir L, Ejaz S, Ashraf M, Qureshi NA, Anjum AA, Iltaf I, Javeed A (2012) Ecotoxicological risks associated with tannery effluent wastewater. Environ Toxicol Pharmacol 34(2):180–191
14. EPA. Prevention pesticides & toxic substances, EPA – 738 – F- 98e00, United States Environmental Protection Agency, 1998. Ortho-phenylphenol (OPP) & sodium ortho-phenylphenate (SOPP) risk characterization document, 2007, Dieteray Exposure Health Assessment Section, Medical Toxicology Branch, Department of Pesticide Regulation
15. Louis ED, Jurewicz EC (2003) Applegate. In: Factor-Litvak P, Parides M, Andrews L, Todd A (eds) Association between essential tremor and blood lead concentration. Environ Health Perspect 285:915–920
16. Hoekstra AY (2017) Water footprint assessment: evolvement of a new research field. Water Resour Manage 31:3061–3081
17. OECD (2012) Environmental outlook to 2050: the consequences of inaction. Organization for Economic Co-Operation and Development, Paris, France
18. Gu Y, Xu J, Keller AA, Yuan D, Li Y, Zhang B, Weng Q, Xiaolei Z, Deng P, Wang H, Li F (2015) Calculation of water footprint of the iron and steel industry: a case study in Eastern China. J Cle Prod, 1–8
19. Gajdzik B, Gawlik R, Skoczypiec S (2018) Forecasting-scenario-heuristic method proposal for assessment of feasibility of steel production scenarios in Poland—Managerial implications for production engineering. Arch Civ Mech Eng 18:1651–1660
20. Gerbens-Leenes PW, Hoekstra AY, Bosman R (2018) The blue and grey water footprint of construction materials: steel, cement and glass. Water Resour Ind 19:1–12

21. Tong Y, Cai J, Zhang Q, Gao C, Wang L, Li P, Hu S, Liu C, He Z, Yang J (2019) Life cycle water use and wastewater discharge of steel production based on material-energy-water flows: a case study in China. J Clean Prod 241:118410
22. Rebitzer G, Ekvall T, Frischnecht R, Hunkeler D, Norris G, Rydberg T, Schmidt WP, Suh S, Weidema BP, Pennington DW (2004) Life cycle assessment part 1: framework, goal and scope definition, inventory analysis, and applications. Environ Int 30:701–720
23. Nezamoleslami R, Hosseinian SM (2020) Data needed for assessing water footprint of steel production. Data Brief 30:105461
24. Ma X, Ye L, Qi C, Yang D, Shen X, Hong J (2018) Life cycle assessment and water footprint evaluation of crude steel productin: a case study in China. J Environ Manag 224:10–19

Employing Input-Output Model to Assess the Water Footprint of Energy System

Li Chai, Aixi Han, Xianglin Yan, and Sai Ma

Abstract The vigorous development of the energy system has imposed a huge burden on the environment, especially for the water environment. On the one hand, the production of energy products requires a large amount of water resources. On the other hand, the environmental pollution caused by the energy system requires a large amount of dilution water. In order to quantify and assess this environmental pressure brought by the energy system, an optimized model is necessary. In this chapter, we start from the concept of water footprint to explain the significant impact of the energy system on the environment. Then we compare the two water footprint evaluation ideas (one is top-down, and another is bottom-up), and led to a more commonly used method, the input–output model. In addition to introducing the basic principles and development process of this model, we also use three case studies to specifically reveal the application of this method in energy and water environment accounting. We hope that our discussion can help readers better understand the function and research results of input–output models in the field of environmental assessment, thereby promoting the further application and deepening of input–output models in the environmental field.

Keywords Input–output model · Input–output analysis · Water footprint · Water pollution · Energy production · Water–energy nexus · Life cycle assessment

1 Introduction: Water Crisis Induced by Energy System

Along with economic development, ecological resources are also deteriorating, and the threat of water resources crisis to humanity is also increasing. Nobel Prize winner Richard Smalley used to point out that in the next 50 years, water-related issues have become the second largest issue that threatens humanity after energy. Water-related matters, including water shortages and water pollution, will severely restrict a country or region's economic and social development. On March 22, 2015, the 23rd World

L. Chai (✉) · A. Han · X. Yan · S. Ma
International College Beijing, China Agricultural University, Beijing, China
e-mail: chaili2005@163.com; chaili@cau.edu.cn

© The Author(s), under exclusive license to Springer   157
Nature Singapore Pte Ltd. 2021
S. S. Muthu (ed.), *Water Footprint*, Environmental Footprints and Eco-design
of Products and Processes, https://doi.org/10.1007/978-981-33-4377-1_6

Water Day, the United Nations report warned that as the global economy grows year by year, humanity may face a water shortage crisis in 2030. From the perspective of foreign and domestic development trends and subject development prospects, the water resources system's sustainable development and management are one of the world's water resources research and international water science research. Due to social development, economic improvement, and environmental changes, water resources systems and related systems have become more and more complex. The concept of green GDP has gradually penetrated industrial production, and the connotation of sustainable development has also been deepened and expanded. Therefore, the rational planning of water resources and effective allocation of water consumption must be coordinated and harmoniously developed with the city's comprehensive development plan, and the development plan should carry out from the scope of the entire city system. Only in this way can the water resources planning be positioned in a proper place, so as to better serve the overall sustainable development of society, economy, and ecological environment.

With the rapid development of industrialization, water and energy have become the main constraints for urban development, because energy supply (such as coal mining, power generation, etc.) directly or indirectly consumes a lot of water and produces a lot of wastewater [18]. The concept of "water-energy nexus" can be introduced as a useful metaphor to study the interdependence of energy and water, and the conversion process embedded in interwoven multidisciplinary chains on multiple scales [4]. At present, the quantitative analysis of the water–energy relationship mainly focuses on the parallel relationship between them: how much water is used for energy supply and what is the energy consumption of water [10, 20, 21]. In these studies, material flow analysis or life cycle assessment is usually used to determine the leverage point of the relevant system by evaluating the path and demand of materials or commodities through the use of a complete supply chain of energy and water [22, 33, 37, 38, 40]. Besides those methods, the input–output (IO) analysis can be used to evaluate the indirect flow of water in addition to the direct flow to illustrate the amount of money needed to produce goods and services based on sector interaction and exchange in a complex system [2, 43, 44].

2 Water Footprint

2.1 Water Footprint Assessment

Water footprint was proposed by Hoekstra at an international expert meeting on virtual water trade in December 2002 in Delft, the Netherlands [14]. Water footprint (WF) of an individual, community, or business is defined as the total volume of freshwater used to produce goods and services consumed by the individual, community, or business. With the increasingly serious problem of water shortage, the use of water resources and the work to solve the problem of water shortage are increasing. The field of water footprint assessment has also emerged as it meets the development needs of consumption and trade [16].

The WF is a measure of consumptive and degradative freshwater use, as Hoekstra defined. The consumptive WF includes a green component, which refers to the consumption of rainwater, and a blue component, which refers to the consumption of surface- or groundwater [15]. The consumption relates to the loss of available surface water and groundwater in the basin. When water evaporates, flowing back out of the basin, sinks into the sea, or is incorporated into products, there appears water loss. The inclusion of the green WF enables the broadening of water resources' perspective beyond the historical focus of water engineers on blue water [8]. The degradative WF, the so-called gray WF, represents the volume of water required to assimilate pollutants entering freshwater bodies [15, 42], which is developed from the concept of dilution water requirement earlier applied by Postel et al. [32].

Water footprint assessment is an analytical tool that can help understand the impact of human activities and products on water shortages and pollution and provide corresponding solutions to ensure the sustainable use of freshwater by human activities and products [31]. Before conducting a water footprint assessment, it is necessary to set reasonable targets and scopes according to the research purpose, collect and calculate data, and carry out a sustainable assessment of water footprint on this basis, and propose targeted response plans.

The water footprint assessment mainly includes four stages.

(1) Set goals and scopes;
(2) Account water footprint;
(3) Sustainable evaluation of water footprint;
(4) Develop a water footprint response plan.

2.2 Water Footprint Assessment Based on Bottom-Up Method (LCA)

So far, the calculation methods of water footprint mainly include two kinds. One is a bottom-up calculation method, in which water is "embedded" in product output, and the product virtual water trade flow analysis is carried out with the help of a water footprint model. Its advantage is that it is flexible and intuitive, and data levels can be selected. However, because it is a data-intensive analysis method, it is not suitable for the national water footprint and is prone to cut off errors. The assessment based on bottom-up method, starting from the perspective of technology and production process, comprehensively considers the production of the product and the requirements in the production process of multi-level raw materials for the sum of all water resources. This method is widely used in the early and mid-term of water footprint research where scholars have adopted this method of life cycle analysis in a series of studies to analyze the water footprint or the issue of virtual water hidden in a certain industry [23, 29, 36]. In the process of analyzing the water footprint of a company or product, especially agricultural products, bottom-up can achieve a more complete and detailed analysis due to its simple and intuitive characteristics.

According to different research objects and purposes, the types of water footprints selected for accounting are also different. The Water Footprint Assessment Manual provides seven types of water footprint [17], including product water footprint, boiler water footprint, and corporate water footprint. Simultaneously, the Manual introduces different calculation methods for different types of water footprints and accounting purposes.

The bottom-up method is suitable for simple early analysis of the life cycle. Once applied to the complex industry analysis, this method shows some shortcomings. First of all, its data acquisition is very challenging. Taking industry as an example, industrial products' production process involves multiple levels of raw material processing, and the production of each level of raw material generates water loss, which is very difficult for data statistics. Second, each industry, region, or country has tens of thousands of products produced. The calculation of the industry's water footprint, region, and country using the bottom-up method must include all products, which is very difficult to operate [19].

2.3   Water Footprint Assessment Based on Top-Down Method (IO Model)

Another way to calculate the water footprint is a top-down accounting method, in which water is "embedded" in the value of each sector of the economy. With the help of the IO model, construct a water resource IO table, conduct virtual water volume research between departments and calculate the water footprint of various products. This method was proposed by the scholar Leontief in the 1930s. The method links the producers and consumers in the entire economic activity through the links of production relations between departments. At the same time, by showing the details of the flow of products between departments, it is possible to trace the water consumption of products in the entire industrial chain [27]. In recent years of research, IO methods have been widely used in the field of environmental footprint assessment due to their simplicity and low data requirements.

3   Input–Output (IO) Model

3.1   The Emergence of IO Method

As early as the 1930s, Leontief, the American economist, has established IO analysis after accepting Walras' general equilibrium theory and Marx's reproduction theory in theory [27, 30]. The IO analysis method mainly includes the IO table and

IO model. The IO model can be used to explore and analyze the interdependence between industries in a region. When the final demand for certain sector changes, the output value change of each industrial sector is estimated [41]. This model mathematically expresses the balance between production and distribution and use in various sectors of the national economy, or the balance between production and consumption in various sectors of the national economy, as reflected in the IO table, and establishes the corresponding IO After the mathematical model, the internal connection between the various sectors of the national economy and the various links of social reproduction can be revealed through computer operations.

3.2 The Features of IO Method

The IO analysis method is an economic quantitative analysis method that puts "input" and "output" together for analysis and research. Generally speaking, "input" refers to the consumption and use of various elements required by the industrial sector in the production process, such as the consumption and use of material products, the consumption and use of labor, and the consumption and acceptance of various production resources or productive services, etc. [6, 39]. The input here includes material production and labor activities. The "output" here refers to the distribution and use of the production results of the industrial sector, that is, the physical movement of the material product or the object of the service, which includes material products or various services [9].

The IO method regards the national economy as a huge economic system composed of systems:

First of all, from a horizontal perspective, the national economy is composed of various departments—goods production sector subsystems (e.g., industry, construction, etc.) and service production sector subsystems (e.g., transportation, life services, finance, and insurance, etc.). Each of those subsystems is composed of several lower level subsystems, and these subsystems can be further subdivided. Subsystems at the same level are interdependent and restrict each other, and there are intricate horizontal connections.

Second, from a vertical perspective, the national economy is composed of various economic activity reproduction subsystems, and each subsystem can be decomposed into more detailed subsystems. First of all, production, distribution, exchange, and consumption are four subsystems where production is the starting point, consumption is the endpoint, and distribution and exchange are the links connecting production and consumption. Production determines consumption, and consumption again gives rise to opportunities for reproduction, therefore, the national economy operate steadily under this interaction. The national economic system has a clear goal: to continuously improve the level of material and technology, produce high-quality, large-volume products, and provide high-quality services to meet the increasing material and cultural needs of the society and the people. The national economic system is an open system. It needs to vigorously develop international connections, continuously introduce advanced technologies, develop foreign trade, export high-quality products to obtain foreign exchange, and import production factors and products needed by the economy. Although the various systems of the national economy are interrelated and have the same goals, in the process of coordinating production, due to limited resource constraints, conflicts in resource allocation cannot be avoided. IO analysis is like a chessboard, all parts are included in a chess game to facilitate overall coordination and overall planning.

3.3 Applying IO Model to Perform Environmental Analysis

Traditional IO accounting is based on the analysis of a purely economic system, without considering the resources and environmental systems closely related to economic activities. In the late 1960s, as resource and environmental problems became more and more serious, scholars in economics began to pay attention to resource and environmental problems, trying to apply IO models to the study of the correlation between economic behavior and resources and environment. Many IO technologies researchers, such as Richard Stone, H. den Hartog, Wassily Daniel Ford, A. Houweling, W. Leontief, J. E. Meade, etc., were the first to use IO analysis techniques to study resource and environmental issues. In the 1970s, W. Leontief was commissioned by the United Nations to use IO technology to study the problems of world economic development and environmental pollution and wrote The Future of The World Economy. In 1972, H. O. Carter and others of the University of California suggested to use the inter-regional IO model to study the use and distribution of the Colorado River in California and Arizona. At the International Pollution Symposium held in Tokyo, Japan in March 1970, Leontief published the paper "Environmental Repercussions and the Economic Structure: An Input-Output Approach." The paper expanded the IO table to include pollutant elimination in the vertical input and pollutant generation in the horizontal direction [24, 25]. This model can be used to analyze the impact of limiting pollution on the industrial structure and price structure, or the cost to be paid by the social economy in order to meet certain environmental standards, so it is forward-looking and innovative. In 1973, Leontief wrote in his paper National Income, Economic Structure, and Environmental Externalities further expands the flow of ordinary goods and services among various departments in the IO table and includes the generation and elimination of pollutants [26].

The energy crisis that emerged in the 1970s promoted the study of considering energy use and energy-related pollutants in IO models. In 1974, Hartog and Houweling considered industrial pollutants in the IO model for the first time [11]. Cumberland and Stram introduced this extended model into the economic policy analysis of environmental constraints [5]. In the 1980s, Hettelingh added different types of energy conversion matrices to the IO model and analyzed the impact of electricity, oil, coal, and other energy components on the environment and economy, and further used the IO method through combining the field of environment and the economy [13]. At the same time, in view of the special relationship between water resources and humans, some scholars in the United States and other countries have done water IO analysis. For example, Carter and others use inter-regional IO models to study the use of Colorado River water in California and Arizona [1]. At the Twelfth International Conference on IO Technology [35], Hynd Bouhia proposed a water IO model and a method for calculating water prices using IO. German scientist Torsley suggested using a generalized IO model to study economic development and pollution control, combining IO technology with mathematics for planning and planning, and calculating the shadow price of water resources and the shadow price of wastewater discharge and treatment, respectively.

In 1993, the United Nations established SEEA (The System of Integrated Environmental and Economic Accounting) as a subsidiary account of SNA and published the SEEA operation manual in 2000, which elaborated on forest resources, land resources, underground assets, water resources, and air pollution. The accounting method describes the environmentally adjusted aggregate indicators and discusses how to apply comprehensive environmental and economic accounting data to economic and environmental policies, evaluate economic behavior, identify environmental problems, evaluate and revise policies, etc.

In 2007, Hawkins and others published the paper A Mixed-Unit Input-Output Model for Environmental Life-Cycle Assessment and Material Flow Analysis [12]. This article combines the material flow analysis model and the economic IO model to create a mixed unit IO model to better track the economic transactions and material flows of the entire economy related to production changes. The resulting model provides the capabilities of logistics and IO models. Through detailed material tracking of the entire supply chain, in response to any currency or material demand, it marks that scholars have officially introduced IO analysis into the environmental analysis. Since then, relevant experts and scholars have done a certain degree of discussion and research on resources, environment, and economic development by improving the traditional IO table.

3.4 Performing an Input-Output Analysis

The subject area of IO analysis is not limited to an empirical research method, it is also a mathematical economics and economic accounting method. It covers the three most important elements of economic analysis in its relatively independent system: theory, data, and methods. These three elements are not independent but closely related. IO analysis method is a kind of economic mathematical model analysis technology, and IO model is its manifestation. The IO technical analysis model includes two steps: constructing the IO table and building the mathematical models.

3.5   Constructing the Input-Output Table

The necessary basis for IO analysis is the IO table, which is a data database that reflects in detail the relationship between the key elements of the national economic system. It is also a balance sheet showing the sources of inputs and destinations of outputs of various sectors (or products) of the economic system. The input column of the table is its main column, and the column direction indicates the consumption and use of various input elements, that is, the source of input. The output column is its object column, and the row direction column indicates where the product is distributed and used after it is produced, that is, where the output goes. The intermediate input and intermediate output of the main column and the object column are staggered horizontally and vertically to form the first quadrant of the table, which is the core part of the IO table. Since the intermediate input items and intermediate output items are divided into n sectors or products, the first quadrant constitutes a table like a chessboard.

The main column of the IO table extends from the intermediate input to the vertical direction; the material form input elements represented by the intermediate input are extended to a wider range of other forms, such as fixed assets, labor, etc., that is, the first quadrant of the table is downward The extension constitutes the third quadrant; the middle product in the guest column of the table expands to the right to other uses beyond the production process of the year, such as investment, consumption, export, etc., that is, the first quadrant of the table expands to the right to constitute the second quadrant. The entire IO table is centered on the first quadrant, which summarizes column-oriented input and row-oriented output and is used to describe the quantitative dependence between various departments or products of the economic system. Simplify the IO table of the national economy as shown in Table 1.

Table 1 An example of Input–Output Table. Note: x_{ij} is the intermediate input from sector i to sector j; U_i is the sum of intermediate outputs of sector i. y_i is the final demand of sector i; in these, C_i is the domestic consumption of sector I and ME_i is the exports of sector i. MI_i is the imports of sector i. T_j is the sum of intermediate inputs to sector j and d_j is the value added to sector j

Output/Input		Intermediate use					Final demand			Imports	Gross output
		1	2	...	n	Sum	Consumption	Exports	Sum		
Intermediate input	1	x_{11}	x_{12}		x_{1n}	U_1	C_1	ME_1	y_1	MI_1	x_1
	2	x_{21}	x_{22}		x_{2n}	U_2	C_2	ME_2	y_2	MI_2	x_2
	...										
	n	x_{n1}	x_{n2}		x_{nn}	U_n	C_n	ME_n	y_n	MI_n	x_n
	Sum	T_1	T_2		T_n	T	C	ME	Y		X
Value added		d_1	d_2		d_n	D					
Gross input		x_1	x_2		x_3	X					

3.6   Mathematical Models for Input-Output Analysis

3.6.1   Mathematical Model

Using linear algebra theory according to the elements recorded in the IO table can establish IO mathematical models and study the quantitative relationship between various departments. IO models mainly include a row model based on the balanced relationship between the input and output tables and a column model based on the balanced relationship between the input and output tables [7]. Another form of IO analysis technology is the mathematical model of IO analysis, which is based on the IO table. It is a way of expressing quantitative relationships and connections using mathematical operation symbols. The IO mathematical model is based on the quantitative relationship between the economic indicators in the IO table, and the mathematical form of linear equations is used to reflect the input in the economic system. And the relationship between output. Therefore, it is necessary to define and calculate certain key coefficients in order to convert the form of the IO table into a mathematical model. After the model is built, various calculations can be carried out according to the needs of economic analysis, and then the calculation results can be obtained, which can be used for empirical analysis, planning, forecasting, and other tasks.

3.6.2   Parameter Settings

(1) Calculate the direct consumption coefficient matrix and the Leontief inverse matrix

The direct consumption coefficient is one of the basic concepts in the IO analysis technology. It is the coefficient value obtained by dividing the total input of the j-th door in the IO table by the value of the product directly consumed by the i-th department. The formula is as follows:

$$a_{ij} = \frac{x_{ij}}{X_j}$$

where a_{ij} represents the input of department i that department j needs to increase unit output; set A be the direct consumption coefficient square matrix.

For the n products in the IO table of the national economy, there must be $n \times n$ direct consumption coefficients. The matrix of $n \times n$ direct consumption coefficients arranged in the order of the standard is called the direct consumption coefficient matrix and is denoted as A as the following:

$$A = \begin{bmatrix} a_{11} & a_{12} & \cdots & a_{1n} \\ a_{21} & a_{22} & \cdots & a_{2n} \\ \cdots & \cdots & \cdots\cdots \\ a_{n1} & a_{n2} & \cdots & a_{nn} \end{bmatrix}$$

The row vector of the direct consumption coefficient matrix A indicates that the i-th ($i = 1,2, \dots ,n$) product is the direct input amount of each product in the production unit. The column vector of matrix A is expressed as the direct consumption of the j-th ($j = 1,2, \dots n$) product to the various products in the row.

(2) Total consumption factor

The complete consumption coefficient refers to the quantity of products of the i-th department that needs to be completely consumed when each unit of the final product is produced by the department. The direct consumption coefficient reflects the direct relationship between various products. It indicates that a certain product directly consumes various products (including itself) in the production process of a unit product, and it refers to the total product. However, in addition to direct links between products, there are also indirect links, and consumption caused by indirect links is called indirect consumption. The sum of direct connection and indirect connection is called complete connection; the sum of system direct consumption and all indirect consumption is called complete consumption. Generally recorded as

Full contact = Direct contact + All indirect contact

Complete consumption = Direct consumption + All indirect consumption

Complete consumption factor = Direct consumption factor + Indirect consumption factor

The complete consumption coefficient is expressed as b_{ij} and its meaning is the complete consumption of the i-th product by the physical product produced by the j-th product. Its relationship with the direct consumption coefficient is expressed as

$$b_{ij} = a_{ij} + \sum_{k=1}^{n} b_{ik}a_{kj}(i, j = 1, 2, \dots, n)$$

where $\sum_{k=1}^{n} b_{ik}a_{kj}$ represents the sum of all indirect consumption of a unit of the j-th product to the i-th product.

Using matrix operations, b_{ij} can be calculated, and B is expressed as

$$B = \begin{bmatrix} b_{11} & b_{12} & \cdots & b_{1n} \\ b_{21} & b_{22} & \cdots & b_{2n} \\ \cdots & \cdots & \cdots\cdots \\ b_{n1} & b_{n2} & \cdots & b_{nn} \end{bmatrix}$$

3.6.3 Different Types for Input–Output Analysis

The most basic models of IO analysis are the value model and the physical model. They are model types divided according to different measurement units. The physical IO model uses the physical quantity of the product as the unit of measurements, such as kilometers, tons, kilograms, joules, and kilowatt-hours. This model can describe in detail the interdependence of physical products in the production process. It is simple and clear. Its application is based on the real economy. The calculation result will not deviate greatly from the actual situation. It is operable with strong performance and high application value. However, because the measurement unit of the physical IO table has great limitations, the physical model is relatively monotonous and simple in terms of expressing the relationship between economic quantities and is limited to physical products in the field of expression, so it cannot fully reflect all economic activities.

At the same time, the IO model that uses value as the unit of measurement is value-based, such as RMB, US dollars, etc. The value-based model unifies the unit of measurement of products produced by the industrial sector, increases the description space of the IO model, and improves the expressiveness of the model. The value model can perform more comprehensively and systematically and reveal the entire economic system. It shows its irreplaceable advantages in terms of economic structure, economic relations, and economic proportional relations.

IO analysis technology is widely used in economic research and economic management. According to different research subjects, different modeling scenarios, and different research tasks, various IO models continue to appear. On the basis of the two most basic models of physical and value, different types of models are compiled in combination with the research characteristics of related fields and the tasks to be completed. The more commonly used models in actual work are

(1) Regional IO model. The regional model needs to be closely integrated with the characteristics of the region, such as the transfer and transfer of the region must be considered. Regional models are generally more complex than national models, but the main task of regional models is to study major issues such as the economic structure and economic development within the region.

(2) Inter-regional IO model. The focus of the model is the economic exchanges and economic ties between several regions. With the strengthening of the global economic integration trend, countries around the world are paying more and more attention to international economic ties, making the inter-regional model a major research focus.

(3) Enterprise IO model. This is a major feature of my country's application of IO models. It mainly studies important issues such as enterprise internal connections, product structure, product prices, cost accounting, and enterprise development planning.

(4) Departmental IO model. For a department or enterprise (medium and large), such as the energy industry sector or energy industry enterprise, etc., part of the products produced by it are used as the final product of the enterprise (or

department), and part of the products are consumed as intermediate products in the internal production plan of the enterprise (or department). There are complex connections. The enterprise IO model is mainly used to solve the problems of how to arrange the supply of various external purchases, how to optimally arrange the production of the enterprise while ensuring that the market demand is met.

(5) Dynamic IO model. For the static model, it does not include the time factor and only shows the IO relationship in a certain period of time. A dynamic model that introduces time changes and other factors can reflect the relationship between input and output changes over time. The dynamic IO model is an economic dynamic model that focuses on the structural changes in economic sectors. It includes time change factors, focusing on the research and analysis of the quantitative relationship and internal relationship between the total investment and composition of the country's economy during the previous period and the scale and structure of the production of various economic sectors in the subsequent period.

3.7 Basic Assumptions of the Input-Output Model

The IO model is an abstract description of economic phenomena, so it is an economic mathematical model. It can only reflect the main characteristics of economic objects, but it cannot reproduce the prototype of economic objects without omission. Based on this, when building the model, it is necessary to use scientific theory as the basis, discard, or abstract away some secondary and non-essential factors and make reasonable assumptions. At the same time, as a mathematical model related to economics, it also needs to link with actual economic phenomena and conditions and consider some logical conditions that should be met in the process of model solving to ensure the normal application of the model in economic applications.

3.7.1 "Pure Sector" (Same Nature) Assumption

The "pure sector" assumption means that any industrial sector only produces a specific product of the same nature, has a single input structure, and uses only one production technology to produce products. This assumption has the following implications:

(1) All products belonging to the same production department can completely replace each other, or these products themselves can be produced in a strict proportional relationship;
(2) Any production department contains only a single input structure;
(3) There is no substitutability for products between different production departments. In other words, the same product or some similar substitutes cannot be classified in different departments.

The significance of this hypothesis allows each department of production to form a single aggregate that produces a certain product, which enables the model to specifically reflect the different uses of the products of each industrial sector and to explain its use direction according to different uses. And it does not take into account the differences in different production technologies and the substitution calculation of products in the internal production process of the industrial sector, which allows the model to accurately reflect the material loss composition of products in different sectors. Therefore, a one-to-one correspondence is established between the production products and the production departments. With this assumption, the material and technical links between departments can be easily represented by material consumption, which is convenient for calculation and analysis using linear methods.

3.7.2 Stability Assumption of Direct Consumption Coefficient

This assumption means that the direct consumption coefficient aij is stable for a certain period of time. This stability includes two meanings: on the one hand, the direct consumption coefficient will not change with time, that is, within a certain period of time, the production technology level of each department is fixed and unchanged, which discards the improvement of labor productivity and technological progress. Factors; On the other hand, the direct consumption coefficient is fixed among enterprises in the same industrial sector.

3.7.3 Proportionality Assumption

This hypothesis refers to the positive relationship between the input and output of various sectors of the national economy, that is, as the production of products increases, the various consumption (inputs) required increase in the same proportion. This assumption is an extension of the direct consumption coefficient assumption.

4 Case Study 1: Using Input–Output Model to Assess the Water Footprint of China's Coal-Fired Power Generation

In this case study, we introduce how to use IO model to perform a life cycle assessment for the water footprint of coal-fired power generation. More detailed analysis and results can be found in our previous published journal article [3].

4.1 Methodology

Using the idea of quantitative analysis, we studied the water depletion and degradation of the entire life cycle of China's coal-fired power plants.

First, we assess the water footprint during the life cycle and compiled Mixed-Unit Input-Output (MUIO) tables. The first thing to note is that electricity prices are not uniform, so in order to avoid the impact of price effects, this study uses the MUIO model to evaluate CPG water quality. The formula can be written as

$$\sum_{j=1}^{n} WF_j = \sum_{j=1}^{n}\sum_{i=1}^{n} WF_i a_{ij} + \sum_{j=1}^{n} DW_j$$

where WF_j indicates the WF of sector j; a_{ij} indicates the demand of sector i by sector j; DW_j indicates the direct water use coefficient, referring to the direct Withdrawal WF, the direct Blue WF, and the direct water pollutants discharge in sector j. The WF and DW are measured in units of m^3 per kWh (electricity) and m^3 per Yuan (other goods and services).

$$WF = DW * (I - A)^{-1}$$

where WF is a vector, comprised of WFs of all sectors; DW is a direct water use coefficient vector; A is a matrix of technical coefficients of intermediate inputs; I is an identity matrix.

At the same time, we estimate three types of water footprints, namely Withdrawal water footprint, blue water footprint, and gray water footprint. The calculation formula for the three footprints is as follows:

$$DW_{grey,p} = \frac{L_p}{C_{max,p} - C_{nat,p}}$$

where $DW_{grey,p}$ (m^3) indicates the direct gray WF of pollutant p and; L_p is the discharge amount of pollutant p to the environment and is measured in a physical unit of grams; $C_{max,p}$ is the maximum permissible concentration (gram per liter) for pollutant p in the water body. $C_{nat,p}$ (gram per liter) is the concentration of pollutant p in the natural water body, which is usually assumed to be zero.

The gray WF (m^3 per kWh) is determined by the maximum specific gray WF among pollutants:

$$WF_{grey} = maxmium\{WF_{grey,1}, WF_{grey,2}, \cdots, WF_{grey,p}\}$$

By considering the differences in regional water shortages, we can better understand the impact of CPG on regional water resources. The Water Stress Index (WSI) has been used by researchers to study the virtual water contained in electrical energy. Its calculation formula is

$$WSI = \frac{1}{1 + e^{-6.4WTA*(1/0.01-1)}}$$

where WTA* is a modified water withdrawal-to-availability dimensionless indicator that considers precipitation variability.

This research introduces the MUIO model as an extension of the standard currency IO analysis. In the MUIO model, the physical flow is included in the currency transaction by rearranging the table, so there are four units in the table reflecting the links between departments, as follows:

$$MUIO\ Table\ = \begin{pmatrix} \frac{kWh}{kWh} & \frac{kWh}{¥} \\ \frac{¥}{kWh} & \frac{¥}{¥} \end{pmatrix}$$

where the numerator in each sub-matrix is the input sector with a unit of energy or money; the denominator is the output sector with a unit of energy or money.

We compile China's 2012 provincial MUIO table and 2002, 2007, and 2012 national MUIO table for 44 industries. Here, the electricity and heat sectors in the original currency information table are divided into four energy sectors: electricity for clean energy products, hydropower, other electricity (i.e., wind energy, solar energy, etc.), and thermal energy. Separate electricity from heat according to the share of each economic value, and then divide electricity into thermal power, hydropower, and the other three industries based on the proportion of the provincial power generation portfolio.

We use World Electric Power Plants Data Base (WEPP) (Utility Data Institute of Platts Energy InforStore, 2015) to identify the cooling technology of a single power plant and estimate the direct water footprint coefficient of the power industry. Figure 1 shows the data compiled by the researchers. In fact, China generally uses the types of cooling technology, namely air cooling, closed-loop cooling, and open-loop cooling. According to relevant data, the researchers established a national direct water footprint coefficient matrix based on different sources to calculate the direct water footprint coefficients of other departments.

4.2 Results

Table 2 lists the water footprints and water pollutant discharge in the life cycle of China's coal-fired power generation in 2012. The shares of the water footprint from different sectors are shown in Table 3. We have calculated that the life cycle WF of withdrawal WF, blue WF, and gray WF are 35.64 m³/MWh, 2.14 m³/MWh, and 17.67 m³/MWh, respectively. The analysis of the composition of the three water footprints, respectively, shows that

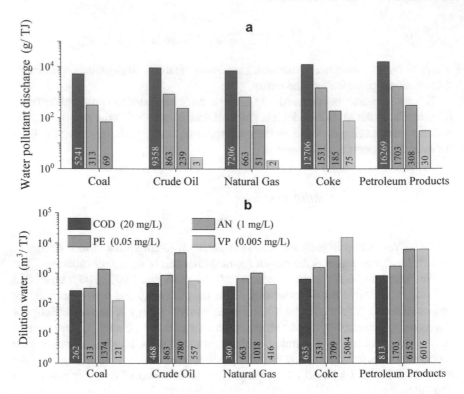

Fig. 1 Life cycle dilution water and water pollutant discharge of fossil fuels per unit of energy. **a** for water pollutant discharge; **b** for dilution water

(1) For withdrawal WF, 96% of it is generated by electricity, and the other part only accounts for about 4%. This is mainly due to the over-utilization of open cooling systems in eastern and southern China, resulting in a water intake of 85.51 m^3/MWh. At the same time, another noteworthy source of water for electricity is the upstream supply chain of the power system, such as the electricity and equipment consumed by the electricity production itself.

(2) For blue WF, its direct water footprint accounts for 53%, and its indirect water footprint (upstream sector consumption) accounts for 47%. In terms of the production sector, electricity still accounts for the largest proportion of blue WF (64%). At the same time, agriculture (21%) and coal (8%) cannot be ignored.

(3) For gray WF, it can be seen from Table 2 that most of the water pollutants are discharged indirectly from the upstream sector (14.88 m^3/MWh), rather than directly discharged to the power station (2.8 m^3/MWh). At the same time, unlike withdrawal WF and blue WF, for gray WF, the coal industry consumes the largest gray water footprint, accounting for 67%. This is because the mine water discharged during coal mining contains a large amount of petroleum pollutant and petroleum pollutant has a very low allowance concentration with only 0.05 mg/L. The gray WF produced by electricity also accounts for a large

Table 2 Water withdrawal, blue water footprint, gray footprint, and water pollutants in the life cycle of China's coal-fired power generation

		Life cycle	Direct	Indirect
Water withdrawal	m^3/MWh	35.64	31.25	4.39
Total Blue WF	m^3/MWh	2.14	1.14	1.00
Gray WF	m^3/MWh	17.67	2.80	14.88
COD	g/MWh	107.42	21.77	85.65
Ammonia nitrogen	g/MWh	8.29	1.43	6.86
Total nitrogen	g/MWh	8.03	0.00	8.03
Total Phosphorus	g/MWh	0.87	0.00	0.87
Petroleum	g/MWh	0.88	0.14	0.74
Volatile phenols	g/MWh	0.03	0.00	0.03
Asenic	mg/MWh	2.96	0.00	2.96
Cadmium	mg/MWh	0.01	0.00	0.01
Cyanide	mg/MWh	0.64	0.00	0.64
Mercury	mg/MWh	1.00	0.00	1.00
Hexavalent Chromium	mg/MWh	2.27	0.00	2.27
Total chromium	mg/MWh	2.50	0.00	2.50
Plumbum	mg/MWh	13.55	10.00	3.55

Table 3 Shares of withdrawal, blue, and gray water footprints from different sectors

Sectors	Water withdrawal (%)	Blue WF (%)	Gray WF (%)
Electricity	93.78	59.26	7.25
Coal	2.55	17.17	62.93
Equipment	1.03	5.90	7.66
Instruments	0.40	2.37	2.78
Industry products	0.95	6.19	14.46
Services	1.19	8.48	4.45
Other sectors	0.09	0.63	0.46

proportion (16%). Pollutants of COD (Chemical Oxygen Demand), petroleum, ammonia nitrogen, volatile phenols, and heavy metals are mainly from industrial sectors, while nitrogen and phosphorus are discharged from the agriculture sector.

Overall, indirect WF accounts for a certain proportion of life cycle WF, especially for gray WF. Therefore, controlling upstream water pollution or water consumption is the key to reducing CPG life cycle WF. At the same time, the power sector has

Table 4 Life cycle Withdrawal, Blue, and Gray water footprints of the coal-fired power plant in China in 2002, 2007, and 2012

		2002	2007	2012
Blue WF	m^3/MWh	2.72	2.60	2.14
Gray WF	m^3/MWh	34.73	49.51	17.67
Water withdrawal	m^3/MWh	40.63	38.44	35.64

more withdrawal WF and blue WF, and the coal sector has a significant impact on gray WF.

In terms of time, we compare China's thermoelectric power's life cycle Withdrawal, Blue, and Gray WFs in the years 2002, 2007, and 2012 to reveal the trend of water footprint over time. As shown in Table 4, Withdrawal WF shows a downward trend year by year where the life cycle water withdrawal has dropped from 41 m^3/MWh in 2002 to 36 m^3/MWh in 2012, a 12% decrease. Blue WF also shows a downward trend year by year, where the life cycle blue water footprint dropped from 2.77 to 2.14 m^3/MWh. If we only compare the start and end years, we find that the gray water footprint dropped by nearly 50% from 34.7 to 17.7 m^3/MWh between 2002 and 2012. In addition, we revealed the reasons for the abnormally high gray water footprint in 2007. Due to the extensive use of chemical fertilizers in the agricultural sector, total nitrogen and total phosphorus peaked in 2007, leading to a higher indirect gray WF. Compared with 2002, oil pollution was effectively alleviated in 2007, and volatile phenolic pollutants dropped to about 40% of the previous level.

From the perspective of the spatial distribution of the withdrawal water footprint and blue water footprint, withdrawal WF shows an obvious trend of gradually decreasing from southeast to northwest. This is due to the wide application of open-loop cooling systems, and the water intake in the southeast region is higher than the national average. Air-cooled systems are commonly used in northwest China, so their water intake and blue water footprints are low. At the same time, the high consumption areas of blue WF are mainly concentrated along the Yangtze River. This is due to the prevalence of closed-loop cooling systems, and the blue water footprint in the northeast and central regions is higher. And in the central region, high Bluewater intensities of coal production also contributed to its high life cycle blue WFs. Later, we use WSIs to adjust Scarce WFs to reveal the impact of thermal power production on regional water shortages. First, high WSIs are mainly located in the eastern and northern coastal areas for the east, the reason is due to dense population and large industrial accumulation, while the lack of water in the north is due to insufficient natural conditions.

4.3 Discussion and Conclusion

Our research results mainly focus on two aspects: one is the quantitative analysis of the three types of water footprints, and the other is the temporal and spatial distribution characteristics of the three types of water footprints.

First, through a mixed unit IO model, the life cycle water consumption of China's CPG is evaluated in terms of water withdrawal, blue, and gray WFs, and they conclude that electricity production dominates the life cycle of water withdrawal. We find that the indirect water footprint occupies a certain proportion in the whole life cycle, so it cannot be ignored. In addition to generating large amounts of energy consumption, thermal power generation will also cause water pollution, which will result in a large gray water footprint. Therefore, the development of water-saving technologies in the electricity sector and water treatment capabilities in the coal sector can help reduce its dependence on water supply. And by improving the water use efficiency of the upstream sector, water consumption can also be effectively reduced.

At the same time, looking at the development and distribution of water footprint from the perspective of time and space, we can correctly judge China's water consumption. In terms of time, the three types of water footprints all show a downward trend, and in particular, the gray water footprint has also declined rapidly after its peak in 2007. Due to different cooling technology applications, the distribution of water footprint consumption has a certain regionality, and the distribution is not even. High consumption areas are roughly distributed in the southeast coast (withdrawal WF) and the Yangtze River Basin (blue WF).

Relevant research results emphasize the geographical changes of CPG and encourage different policy priorities in different regions. As for the choice of cooling technology, we believe that closed-loop technology should be widely used throughout the country due to its water-saving and relatively low cost. Also, seawater is a potential water resource that can be used in its open-loop cooling systems.

5 Case Study 2: Using Input–Output Model to Assess the Water Footprint of China's Fossil Fuels Production

Global energy demand is expected to increase by 48% from 579 billion GJ in 2012 to 860 billion GJ in 2040, posing a severe challenge to limited natural resources such as freshwater for energy production. In this case study, we introduce how to conduct an IO analysis of the gray water footprint of the energy sector. This can help us understand the contribution of the entire society to the life cycle water footprint of fossil energy production. More detailed analysis and results can be found in our previous published journal article [34].

5.1 Methodology

We first determined the water degradation intensity index, dilution water, which means the minimum freshwater requirement for diluting pollutants at a safe concentration. According to this definition, gray water footprint is defined as the largest dilution water among pollutants, and this concept can be used to evaluate key water pollutants. Using this definition, we used the MUIO to calculate the gray water footprint life cycle assessment of the energy sector:

$$LCW = DW*(I - A) - 1$$

where LCW is the water consumption or pollution intensity vector in the whole life cycle production process; DW represents the direct water consumption or pollution intensity vector; I is an identity matrix as defined in the IO model formula; A is a technical coefficient index matrix which is related to intermediate inputs. We need to minus the virtual water flow between different kinds of fossil fuel industries in order to avoid calculating twice. The following equation could be presented after adaption:

$$LCW_{total} = \sum_{i=1}^{5} \left(LCW_i - \sum_{j=1}^{5} W_{i,j} \right)$$

where LCW_{total} is the life cycle virtual water which is required by fossil fuel industries and LCW_i in the formula means the i-th fossil fuel sector's life cycle virtual water. $W_{i,j}$ is the virtual water of fossil fuel sector i contributed by fossil fuel sector j.

We chose the 2012 IO table, and after performing calculation and analysis on the basis of the original IO table, we reorganized the two energy industries of crude oil and natural gas extraction industry and refined oil coking industry into a new form: crude oil, natural gas, coke, and petroleum products. The calculating formula for these four sectors is listed as follows:

$$\begin{cases} a_m P_m + a_n P_n = a_{m\&n}(P_m + P_n) \\ a_m/a_n = a'_n/a'_m \end{cases}$$

In the formula, a_m or a_n represent the direct demand index of each energy sector expressed by sector m or sector n in the research year. Another index a' represents the direct demand index in the reference year and expression $m\&n$ refers to the combined sector of m and n in the original IO table. The total production number is represented by Pm(n) which has a physical unit instead of a monetary unit that could help avoiding effects by inflation.

Then, four kinds of typical pollutants which could be produced from the energy sectors in China: chemical oxygen demand (COD), ammonia nitrogen (AN), petroleum (PE), and volatile phenol (VP) are selected and studied because of the high emission volume and huge damage to the water resource in China. The dilution water for each pollutant is estimated as follows:

$$DW_{dilution} = L/(C_{max} - C_{nat})$$

where $DW_{dilution}$ is the direct dilution water(m^3); L is the discharge amount of the studied water pollutant(g);C_{max} is the maximum allowable concentration of the studied water pollutant set by regulation (g/m^3); and C_{nat} is the concentration of the studied water pollutant in the natural water body (g/m^3). The pollutants of COD, AN, PE, and VP are allowed to have the maximum concentrations of 20, 1, 0.05, and 0.005 mg/L, respectively, based on this standard.

$$LCW_{grey} = max\{LCW_{dilution,1}, LCW_{dilution,2}, LCW_{dilution,3}, LCW_{dilution,4}\}$$

where LCW_{grey} is the gray water footprint of the life cycle; $LCW_{dilution}$ is the life cycle dilution water of each pollutant.

5.2 Results

First of all, it can be seen from Table 5 that comparing the five products, coal consumes and withdraws the lowest amount of water, so end-user coal is superior to other fuels in terms of saving water. Secondly, crude oil and natural gas use more water than coal, and upgraded fuels, including coke and petroleum products, consume more water due to energy loss during the upgrade process.

Second, we identified the pollutants that have the greatest impact on water quality as shown in Fig. 1. We found that first, the water pollution of coke and petroleum products extracted from coal and crude oil is always heavier than that of raw fuel and produces a large amount of VP dilution water. Meanwhile, PE produces a large amount of dilution water for all fossil fuels and becomes a decisive factor in the gray water footprint of coal, crude oil, natural gas, and petroleum products. Another thing worth noting is that although COD has the highest emissions in the fossil energy industry, reaching 5–16 kg/TJ as shown in Fig. 1a, its allowable concentration is

Table 5 Water consumption, water withdrawal, wastewater discharge, and gray water footprint in the life cycle production of fossil fuels (Unit: m^3 per TJ)

Items	Coal	Crude oil	Natural gas	Petroleum products	Coke
Water consumption	39	121	97	91	220
Water withdrawal	142	479	381	456	717
Wastewater discharge	123	358	284	385	498
Gray water footprint	1374	4780	1018	15084	6152

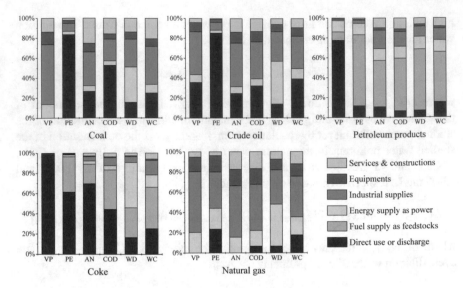

Fig. 2 Life cycle shares of fossil fuel industries, including WC (water consumption), WD (water withdrawal), COD (chemical oxygen demand discharge), AN (ammonia nitrogen discharge), PE (petroleum discharge), and VP (volatile phenols discharge)

relatively high (20 mg/L), so the dilution water demand is relatively small, therefore, COD is not the decisive pollutant of the gray water footprint.

Figure 2 shows the share of different sectors in the life cycle of water consumption and pollution, which can help us identify the contributions of different sectors. We can see that in the life cycle production of coal and crude oil, water is mainly consumed and extracted by industrial sectors, such as power generation and chemical production. The water pollution caused by the coking industry is very serious (40% for AN, 48% for COD, 70% for PE, and 98% for VP). In addition, petroleum products, including gasoline and diesel, are all refined from crude oil. Therefore, in addition to VP pollution, water is mainly used and polluted by the crude oil sector.

When analyzing the performance of water-saving and emission reduction in the fossil energy industry, we first found that the coal industry has the lowest pollutant removal rate, partly because its wastewater treatment costs are low. Therefore, the massive production of coal mine wastewater has brought great cost pressure on wastewater treatment. In the production process for crude oil, natural gas, petroleum products, and coke, up to 7% of the profits are used for wastewater treatment, so its water pollution control performance is above average, that is, the pollutant removal rate and treatment cost are higher than the entire industry. However, it is worth noting that PE and VP are still the most serious pollutants causing water pollution in the fossil fuel industry.

We designed three scenarios (S0, S1, and S2) to investigate the mitigation of water pollution when pollutants exceeding the prescribed standards are removed. S0 refers to the situation where the concentration of water pollutants in the discharged

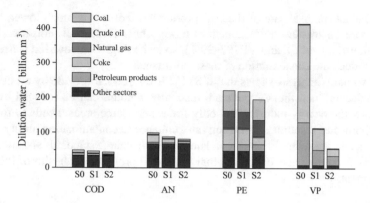

Fig. 3 Life cycle dilution water of China's fossil fuel production pollutants under three scenarios (S0, S1, and S2) in 2020

wastewater remains unchanged and equal to the current average value. For S1 and S2, all excessive pollutants are treated and discharged according to the concentration of the corresponding standards (STD1 and STD2). The results are shown in Fig. 3. First, if all the pollutant emissions of the coking industry meet STD1, the VP emissions can be greatly reduced. After the effective removal of VP reaches the standard, the fossil fuel industry should focus on controlling PE emissions. There is no significant difference between COD and AN discharge in the three scenarios because these two pollutants are mainly discharged from other industrial sectors rather than directly from the fossil fuel sector.

5.3 Discussion and Conclusion

This research has developed a mixed unit IO model to evaluate the life cycle of water degradation during the production process of China's most common fossil fuels (including coal, crude oil, natural gas, coke, and petroleum products.). At the same time, scenario analysis is set up to implement forecasting environmental benefits brought by different pollutant emission standards. Studies have found that in all cases, the amount of gray water produced is much greater than the amount of water consumed and taken. Although the chemical oxygen demand (COD) emissions from fossil fuel production are large, petroleum (PE) and volatile phenol (VP) require more dilution water than COD, so they need to be taken into consideration.

At the same time, PE is the biggest contributor to water degradation caused by primary fossil fuels, and VP pollution is particularly prominent in upgrading fossil fuel production. It is worth noting that the high toxicity of VP makes it need to be focused on, away from residential areas. The main cause of water degradation is the direct discharge of PE and VP, which mainly occur in coal mines, oil fields, refineries, and coking plants, and the upstream sector accounts for only a small proportion.

The pollutant removal rate of the coal production industry is much lower, but the expenditure on wastewater treatment is more. Other fossil fuel industries have a high removal rate of PE and VP (97–99%), so technological innovation is needed to further reduce the concentration of these pollutants.

Our scenario analysis shows that if STD2 is obeyed nationwide by all the fossil fuel producers, then the VP pollutant discharge amount can be reduced by 88%. Therefore, the energy industry, not only for water-scarce areas, needs to focus on improving water resources utilization rate, enhance decontamination ability. At the same time, the state should also formulate stricter discharge standards and strengthen supervision on the basis of the original system to promote the health of the water environment.

6 Case Study 3: Using Input–Output Model to Assess the Water Footprint of Energy Consumption by Chinese Households

In this case study, we introduce how to use the IO model to assess the water footprint of energy from the perspective of household consumption. This can help us to understand how household energy consumption affects water resources. More detailed analysis and results can be found in our previous published journal article [28].

6.1 Methodology

We used the IO analysis method to quantify the life cycle water consumption of Chinese household energy demand from 2002 to 2015 and compared the difference between rural and urban. Then through structural decomposition analysis, the impact of the four driving factors of population, demand, economic structure, and technology is clarified. The related methods are as follows:

First of all, the basic equation for quantifying the life cycle water consumption of household energy consumption can be expressed as

$$W_u = W * (I - A)^{-1} Y_{e,u}$$

$$W_r = W * (I - A)^{-1} Y_{e,r}$$

where W_u and W_r are the life cycle water uses to meet urban and rural household energy consumption $Y_{e,u}$ and $Y_{e,r}$, respectively; $(I - A)^{-1}$ is the Leontief inverse matrix, also called total requirement matrix, that represents the required inputs from each sector to fulfill each sectors' final demands, in which I is an identity matrix and A is the matrix of inter-sector intermediate input coefficients; $W^* = [W_1^*, W_2^*, \ldots$

W_n *] is a row vector of all sector's water intensities, which equals to direct water inputs in each sector dividing the sector's economic output.

Then, household energy Y_e can be decomposed to population P, including urban and rural, and household energy consumption per capita ye (MWh/p) according to the equation: Impact = Population × Affluence × Technology (IPAT) model.

$$W_u = W * \cdot (I - A)^{-1} \cdot P_u \cdot y_{e,u}$$

$$W_r = W * \cdot (I - A)^{-1} \cdot P_r \cdot y_{e,r}$$

where water use efficiency of each economic sectors $W*$ denotes the Technology Effect; total requirements matrix $(I - A)^{-1}$ represents the Structure Effect; P is the Population Effect; and y_e is the Demand Effect.

We use an additive mathematical form to identify four driving factors:

$$\Delta W = W^{*'} + L' + P' + y_e'$$

where $W^{*'}, L', P' and y_e'$ denote the impacts brought by changes of water intensities $W*$, Leontief inverse matrix $(I - A)^{-1}$, population P, and energy consumption per capita y_e, respectively.

$$\Delta W = W^t - W^0 = w^{*t} L^t P^t y_e^t - w^{*0} L^0 P^0 y_e^0$$
$$= \left(w^{*0} + \Delta w^* \right) \left(L^0 + \Delta L \right) \left(P^0 + \Delta P \right) \left(y_e^0 + \Delta y_e \right) - w^{*0} L^0 P^0 y_e^0$$

where superscripts all denote either the start, 0, or the endpoint, t, of the time period $[0, t]$ and Δ represents the changes of corresponding variables during this time period.

An example equation to quantify $w^{*'}$ is as below:

$$w^{*'} = \Delta w^* L^0 P^0 y_e^0 + \frac{1}{2} \Delta w^* (\Delta L P^0 y_e^0 + L^0 \Delta P y_e^0 + L^0 P^0 \Delta y_e)$$
$$+ \frac{1}{3} \Delta w^* (\Delta L P^0 y_e^0 + L^0 \Delta P y_e^0 + L^0 P^0 \Delta y_e) + \frac{1}{4} \Delta w^* \Delta L \Delta P \Delta y_e$$

Similarly, L', P', and y_e' can also be quantified.

For the selection of data, we used four time series IO tables for 2002, 2007, 2012, and 2015 of China's 32 industries provided by China's national statistics. And water use data include both water withdrawal and water consumption. Water withdrawal data are obtained from the Water Resource Bulletins in these 4 years. Then, multiply the water withdrawal data of each department by the water consumption coefficient of that department (taken from the Water Resources Bulletin) to convert it into water consumption.

6.2 Results

First of all, from the perspective of time development, from 2002 to 2015, the life cycle water withdrawal and water consumption of household energy consumption experienced a fluctuating process, but overall it is faced with a slow decrease. Geographically, the life cycle water consumption of urban household energy consumption is about four times that of rural water consumption. At the same time, although the proportion of power and heat water consumption in the whole life cycle is gradually decreasing, it still cannot be ignored. In 2015, the whole life cycle water consumption of power and heat was 5.66 km^3, accounting for 74.6% of the whole household energy consumption and the whole life cycle water consumption. At the same time, since 2002, the life cycle water consumption of household gas and petroleum products has steadily increased. The life cycle water consumption of household coal consumption accounts for the smallest share.

We studied the blood supply of the upstream sector of the final energy demand life cycle water from 2002 to 2015, which can better identify the source of energy consumption. We found that the Agricultural, Husbandry, Forestry and Fishing (AHFF) sector and Oil and Natural Gas Extraction (ONGE) sector contributed the biggest shares among all upstream sectors. In particular, for coal consumption, AHFF accounted for its upstream water intake 70.0% in 2015 and accounted for 84% for its upstream water consumption.

By analyzing the driving factors of changes in water use throughout the life cycle, we found that, first of all, the per capita coal consumption of urban residents continues to decline, while the demand for petroleum products continues to rise. Therefore, the status of coal is being replaced by petroleum products. At the same time, the per capita electricity consumption and heat demand of urban households showed an overall upward trend but declined from 2007 to 2012, which may be related to the global financial crisis. In general, technological advancement can improve the water use efficiency of other sectors and reduce the common benefits of household energy consumption dependent on water.

6.3 Discussion and Conclusion

The improvement of people's living standards is accompanied by the growth of energy consumption. However, this growth has put increasing pressure on the natural environment. A lot of water is used for power and heat, but this proportion is declining, while the proportion of gas consumption and petroleum product consumption is increasing. At the same time, the life cycle water consumption of energy consumption also depends on its upstream sector. We found that AHFF is the largest contributor to upstream water consumption.

For the problems we have found, we suggest using various policy tools to curb the pressure of residents' energy consumption on water. Such as subsidies for energy-saving appliances and popularization of public transportation. At the same time, for thermoelectric power plants that consume a lot of water (they also bring a lot of water pollution and produce a lot of gray water footprint), they need to improve their energy conversion efficiency and improve their cooling technology. Since the life cycle water consumption of energy consumption also depends on its upstream sector, a concerted effort is needed to save water throughout the economy.

References

1. Carter H O, Ireri D (1970) Linkage of California—Arizona input—output model to analyze water transfer pattern. Appl Input Output Anal 11(3):139–168
2. Cazcarro I, Duarte R, Chóliz JS (2013) Multiregional input-output model for the evaluation of spanish water flows. Environ Sci Technol 47:12275–12283
3. Chai L, Liao X et al (2018) Assessing life cycle water use and pollution of coal-fired power generation in china using input-output analysis. Appl Energy 231:951–958
4. Chen B, Chen SQ (2015) Urban metabolism and nexus. Eco. Info 26:1–2
5. Cunmber JH, Stram BN (1976) Emperical application of input—output models to environmental problems. In: Advances in input—output analysis
6. Dietzenbacher E, Velazquez E (2007) Analyzing Andalusian virtual water trade in an input—output framework. Reg Stud 41(2):185–196
7. Dietzenbaeher E, Los B (1998) Structural decompose situation techniques: sense and sensitivity. J Econ Syst Res 10(3):307–324
8. Falkenmark M, Rockström J (2004) Balancing water for humans and nature: the new approach in ecohydrology. Earthscan, London, UK
9. Guan DB, Hubacek K (2007) Assessment of regional trade and virtual water flows in China. Eco Econ 61:159–170
10. Hardberger A (2013) Powering the tap dry: regulatory alternatives for the energy water nexus. Univ Colorado Law Rev 84:529–861
11. Hartog H, Houwcling A (1974) Pollution abatement and the economic structure: empirical results of input output computations for the Netherlands. Occasional No. 1, The Hauage: Central Planning Bureau
12. Hawkins T, Hendrickson C, Higgins C, Matthews HS, Suh S (2007) A mixed-unit input-output model for environmental life-cycle assessment and material flow analysis. Environ Sci Technol 41(3):1024–1031
13. Hettelingh et al (1985) A modeling and information system for environmental policy in the Netherlands. Amst Inst Neth Stud E-85/1. Free University
14. Hoekstra AY (ed) (2003) Virtual water trade: proceedings of the international expert meeting on virtual water trade. Delft, The Netherlands, 12–13 December 2002. In: Value of water research report series, No 12. IHE, Delft, The Netherlands
15. Hoekstra AY et al (2011) The water footprint assessment manual: setting the global standard. Earthscan, London, UK
16. Hoekstra AY (2016) A critique on the water-scarcity weighted water footprint in LCA. Ecol Indic 66:564–573
17. Hoekstra Arjen Y (2011) The water footprint assessment manual: setting the global standard. Earthscan, London; Washington, DC
18. Hoff H (2011) Understanding the nexus. Stockholm Environment Institute, Stockholm

19. Jefferies D et al (2012) Water footprint and life cycle assessment as approaches to assess potential impacts of products on water consumption. Key learning points from pilot studies on tea and margarine. J Clean Prod 33:155–166
20. Kenway SJ, Lant PA, Priestley A, Daniels P (2011) The connection between water and energy in cities: a review. Water Sci Technol 63(9):1983–1990
21. Kenway SJ, Mcmahon J, Elmer V, Conrad S, Rosenblum J (2013) Managing waterrelated energy in future cities—a research and policy roadmap. J Water Clim Change 4(3):161–175
22. King CW, Stillwell AS, Twomey KM, Webber ME (2013) Coherence between water and energy policies. Nat Resour J 53(1):117–215
23. Koehler A (2008) Water use in LCA: managing the planet's freshwater resources. Int J Life Cycle Assess 13(6):451–455
24. Leontief W (1970a) Environmental repercussions and the economic structure: an input—output approach. Rev Econ Stat 52:262–277. Reprinted in Leontief W (1977) Essays in economics, II. Basil Blackwell, Oxford; Leontief W (1986) Input—output economics, 2nd ed. Oxford University Press, Oxford
25. Leontief W (1970b) The dynamic inverse. In: Carter AP, Brody A (eds) Contributions to input—output analysis. North-Holland, Amsterdam, pp 17–46. Reprinted in Leontief W (1977) Essays in economics, II. Basil Blackwell, Oxford; Leontief W (1986) Input—output economics, 2nd edn, Oxford University Press, Oxford
26. Leontief W (1973) National income, economic structure and environmental externalities. In: Studies in income and wealth, vol 38. In: Moss M (ed) The measurement of economic and social performance. National Bureau of Economic Research, New York, pp 565–576. Reprinted in Leontief W (1977) Essays in economics, II. Basil Blackwell, Oxford; Leontief W (1986) Input—output economics, 2nd edn. Oxford University Press, Oxford
27. Leontief W (1966) Input-output economics. Oxford University Press
28. Liao X, Chai L et al (2019) Life-cycle water uses for energy consumption of chinese households from 2002 to 2015. J Environ Manag 231:989–995
29. Milà i Canals L et al (2009) Assessing freshwater use impacts in LCA: Part I—Inven-tory modelling and characterisation factors for the main impact pathways. Int J Life Cycle Assess 14(1):28–42
30. Munksgaard J, Wier M, Lenzen M, Dey C (2005) Using input–output analysis to measure the environmental pressure of consumption at different spatial levels. J Ind Ecol 9:169–186
31. Namchancharoen T (2015) The carbon and water footprint assessment of cassava-based bioethanol production in Thailand. In: International institute of chemical, biological and environmental engineering. Proceedings of international conference on biological, environment and food engineering (BEFE-2015, Singapore). International Institute of Chemical, Biological and Environmental Engineering, 6
32. Postel SL, Daily GC, Ehrlich PR (1996) Human appropriation of renewable freshwater. Science 271:785–788
33. Stillwell AS, King CW, Webber ME, Duncan IJ, Hardberger A (2009) Energy-water nexus in texas; environmental defense fund. University of Texas at Austin, Austin, TX, USA
34. Su Y et al (2019) Water degradation by China's fossil fuels production: a life cycle assessment based on an input–output model. Sustainability (Basel, Switzerland) 11(15):4130
35. The Report of the Twelfth International Conference on Input- Output Techniques at New York 9(1), 50–52, 1999
36. Pacetti T, Lombardi L, Federici G (2015) Water–energy Nexus: a case of biogas production from energy crops evaluated by Water Footprint and Life Cycle Assessment (LCA) methods. J Clean Prod 084:278–291
37. U.S. Department of Energy. The water-energy nexus: challenges and opportunities. Washington DC: Department of Energy; 2014
38. Venkatesh G, Chan A, Brattebo H (2014) Understanding the water–energy–carbon nexus in urban water utilities: comparison of four city case studies and the relevant influencing factors. Energy 75:153–166

39. Wang Saige, Chen Bin (2019) Accounting framework of energy-water Nexus Technologies based on 3 scope hybrid life cycle analysis. Energy Procedia 158:4104–4108
40. Wang Saige, Chen Bin (2016) Energy–water nexus of urban agglomeration based on multiregional input–output tables and ecological network analysis: a case study of the Beijing–Tianjin–Hebei region. Appl Energy 178:773–783
41. Wassily L (1986) Input-output economics. Oxford University Press, Oxford
42. Xinchun C, Mengyang W, Rui S, La Z, Dan C, Guangcheng S, Xiangping G, Weiguang W, Shuhai T (2018) Water footprint assessment for crop production based on field measurements: a case study of irrigated paddy rice in East China. Sci Total Environ: 610–611
43. Zhang Y, Zheng HM, Yang ZF, Li YX, Liu GY, Su MR et al (2015) Urban energy flow processes in the Beijing–Tianjin–Hebei (Jing–Jin–Ji) urban agglomeration: combining multi–regional input–output tables with ecological network analysis. J Clean Prod 114:243–256
44. Zhao X, Chen B, Yang ZF (2009) National water footprint in an input–output framework—a case study of China 2002. Ecol Model 220(2):245–253

Printed in the United States
by Baker & Taylor Publisher Services